I0427467

# Basic Math Workbook for Adults

## Addition, Subtraction, Multiplication and Division

2000 exercises with Answer key included

By:
ARC's Education

Hey there! Ready to conquer math anxiety and boost your skills? Dive into our basic workbook for adults, featuring 2000 exercises divided into 100 days of timed tests.

Track your progress with built-in scorekeeping and find answers included for easy reference.

If you've found our workbook helpful, share your experience on Amazon and inspire others to join the journey.

At ARC's Education, we're committed to making math accessible and enjoyable for all. Let's make math a joyous adventure together!

# Contents

 Name: _____      Date: _____

Time Taken: ___:___     Score: ____ / 20

**Day 1**

1)
```
   55
+  18
_____
```

2)
```
   68
+  38
_____
```

3)
```
   86
+  67
_____
```

4)
```
   91
+  76
_____
```

5)
```
    5
+  84
_____
```

6)
```
   31
+  64
_____
```

7)
```
   62
+   5
_____
```

8)
```
   64
+  72
_____
```

9)
```
   92
+   1
_____
```

10)
```
   72
+  65
_____
```

11)
```
   56
+  66
_____
```

12)
```
   38
+  99
_____
```

13)
```
   19
+   3
_____
```

14)
```
   43
+  65
_____
```

15)
```
   79
+  56
_____
```

16)
```
   34
+  43
_____
```

17)
```
   32
+  75
_____
```

18)
```
   79
+  65
_____
```

19)
```
   60
+  40
_____
```

20)
```
   44
+   8
_____
```

  Let's start our day with a dash of addition!

 **Name:** _____

 **Date:** _____

**Day 2** **Time Taken:** ___:___

 **Score:** / **20**

1)
$$\begin{array}{r} 82 \\ + 67 \\ \hline \end{array}$$

2)
$$\begin{array}{r} 4 \\ + 13 \\ \hline \end{array}$$

3)
$$\begin{array}{r} 64 \\ + 8 \\ \hline \end{array}$$

4)
$$\begin{array}{r} 17 \\ + 63 \\ \hline \end{array}$$

5)
$$\begin{array}{r} 6 \\ + 82 \\ \hline \end{array}$$

6)
$$\begin{array}{r} 79 \\ + 91 \\ \hline \end{array}$$

7)
$$\begin{array}{r} 92 \\ + 90 \\ \hline \end{array}$$

8)
$$\begin{array}{r} 47 \\ + 65 \\ \hline \end{array}$$

9)
$$\begin{array}{r} 45 \\ + 79 \\ \hline \end{array}$$

10)
$$\begin{array}{r} 31 \\ + 66 \\ \hline \end{array}$$

11)
$$\begin{array}{r} 31 \\ + 73 \\ \hline \end{array}$$

12)
$$\begin{array}{r} 73 \\ + 61 \\ \hline \end{array}$$

13)
$$\begin{array}{r} 33 \\ + 34 \\ \hline \end{array}$$

14)
$$\begin{array}{r} 85 \\ + 17 \\ \hline \end{array}$$

15)
$$\begin{array}{r} 68 \\ + 4 \\ \hline \end{array}$$

16)
$$\begin{array}{r} 25 \\ + 10 \\ \hline \end{array}$$

17)
$$\begin{array}{r} 84 \\ + 6 \\ \hline \end{array}$$

18)
$$\begin{array}{r} 10 \\ + 97 \\ \hline \end{array}$$

19)
$$\begin{array}{r} 78 \\ + 2 \\ \hline \end{array}$$

20)
$$\begin{array}{r} 80 \\ + 85 \\ \hline \end{array}$$

 Addition is the name of the game today. Let's play!

1)
$$67 + 3$$

2)
$$26 + 99$$

3)
$$49 + 18$$

4)
$$85 + 84$$

5)
$$9 + 41$$

6)
$$87 + 21$$

7)
$$66 + 41$$

8)
$$24 + 76$$

9)
$$76 + 69$$

10)
$$25 + 46$$

11)
$$98 + 10$$

12)
$$54 + 81$$

13)
$$34 + 49$$

14)
$$94 + 14$$

15)
$$34 + 78$$

16)
$$61 + 66$$

17)
$$51 + 41$$

18)
$$34 + 83$$

19)
$$69 + 7$$

20)
$$19 + 51$$

  Time to add up some fun! Let's get calculating!

 **Name:** _____   **Date:** _____

**Time Taken:** ___:___   **Score:** ___ / **20**

**Day 4**

1)
$$\begin{array}{r} 21 \\ + 73 \\ \hline \end{array}$$

2)
$$\begin{array}{r} 90 \\ + 47 \\ \hline \end{array}$$

3)
$$\begin{array}{r} 83 \\ + 45 \\ \hline \end{array}$$

4)
$$\begin{array}{r} 67 \\ + 63 \\ \hline \end{array}$$

5)
$$\begin{array}{r} 24 \\ + 53 \\ \hline \end{array}$$

6)
$$\begin{array}{r} 14 \\ + 20 \\ \hline \end{array}$$

7)
$$\begin{array}{r} 50 \\ + 56 \\ \hline \end{array}$$

8)
$$\begin{array}{r} 35 \\ + 2 \\ \hline \end{array}$$

9)
$$\begin{array}{r} 90 \\ + 93 \\ \hline \end{array}$$

10)
$$\begin{array}{r} 31 \\ + 64 \\ \hline \end{array}$$

11)
$$\begin{array}{r} 47 \\ + 15 \\ \hline \end{array}$$

12)
$$\begin{array}{r} 93 \\ + 55 \\ \hline \end{array}$$

13)
$$\begin{array}{r} 67 \\ + 24 \\ \hline \end{array}$$

14)
$$\begin{array}{r} 1 \\ + 70 \\ \hline \end{array}$$

15)
$$\begin{array}{r} 95 \\ + 56 \\ \hline \end{array}$$

16)
$$\begin{array}{r} 87 \\ + 76 \\ \hline \end{array}$$

17)
$$\begin{array}{r} 93 \\ + 6 \\ \hline \end{array}$$

18)
$$\begin{array}{r} 47 \\ + 1 \\ \hline \end{array}$$

19)
$$\begin{array}{r} 67 \\ + 24 \\ \hline \end{array}$$

20)
$$\begin{array}{r} 4 \\ + 27 \\ \hline \end{array}$$

Let's make math magic happen with addition!

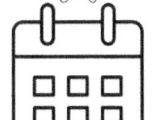 
1)
```
   34
+  98
_____
```

2)
```
   77
+  34
_____
```

3)
```
   57
+  64
_____
```

4)
```
   98
+  84
_____
```

5)
```
   78
+  65
_____
```

6)
```
   69
+  53
_____
```

7)
```
   80
+  56
_____
```

8)
```
   43
+  60
_____
```

9)
```
    5
+  58
_____
```

10)
```
   28
+  11
_____
```

11)
```
   11
+  28
_____
```

12)
```
   36
+  82
_____
```

13)
```
   53
+  89
_____
```

14)
```
   32
+  84
_____
```

15)
```
   51
+  65
_____
```

16)
```
   55
+  77
_____
```

17)
```
   57
+  86
_____
```

18)
```
   71
+  88
_____
```

19)
```
   89
+  78
_____
```

20)
```
   17
+   1
_____
```

  Addition adventures await! Let's embark on our journey!

1)  
$$25 + 62$$

2)  
$$68 + 32$$

3)  
$$7 + 49$$

4)  
$$16 + 21$$

5)  
$$19 + 23$$

6)  
$$82 + 38$$

7)  
$$80 + 20$$

8)  
$$63 + 4$$

9)  
$$42 + 13$$

10)  
$$57 + 4$$

11)  
$$10 + 51$$

12)  
$$20 + 23$$

13)  
$$92 + 64$$

14)  
$$98 + 5$$

15)  
$$12 + 96$$

16)  
$$44 + 54$$

17)  
$$28 + 17$$

18)  
$$55 + 19$$

19)  
$$5 + 37$$

20)  
$$58 + 58$$

Adding numbers is our mission. Let's conquer it together!

 **Name:** _____

 **Date:** _____

**Day 7**

 **Time Taken:** ___:___

 **Score:**    /  **20**

1)
```
   90
 + 43
 ____
```

2)
```
   71
 + 89
 ____
```

3)
```
   67
 + 33
 ____
```

4)
```
   94
 + 73
 ____
```

5)
```
   63
 + 53
 ____
```

6)
```
   95
 + 55
 ____
```

7)
```
   56
 + 75
 ____
```

8)
```
   33
 + 33
 ____
```

9)
```
   51
 + 63
 ____
```

10)
```
   12
 + 33
 ____
```

11)
```
   31
 + 67
 ____
```

12)
```
   57
 +  3
 ____
```

13)
```
   36
 +  3
 ____
```

14)
```
   83
 + 76
 ____
```

15)
```
   89
 + 11
 ____
```

16)
```
   36
 + 58
 ____
```

17)
```
   71
 + 84
 ____
```

18)
```
   39
 + 53
 ____
```

19)
```
   85
 + 42
 ____
```

20)
```
   24
 + 48
 ____
```

10

  Let's kick off our math practice with some addition action!

 **Name:** _____

 **Date:** _____

**Day 8**  **Time Taken:** ___:___   **Score:** ___ / 20

1)
$$
\begin{array}{r} 37 \\ + 53 \\ \hline \end{array}
$$

2)
$$
\begin{array}{r} 70 \\ + 36 \\ \hline \end{array}
$$

3)
$$
\begin{array}{r} 94 \\ + 55 \\ \hline \end{array}
$$

4)
$$
\begin{array}{r} 43 \\ + 31 \\ \hline \end{array}
$$

5)
$$
\begin{array}{r} 90 \\ + 70 \\ \hline \end{array}
$$

6)
$$
\begin{array}{r} 13 \\ + 62 \\ \hline \end{array}
$$

7)
$$
\begin{array}{r} 19 \\ + 37 \\ \hline \end{array}
$$

8)
$$
\begin{array}{r} 44 \\ + 50 \\ \hline \end{array}
$$

9)
$$
\begin{array}{r} 75 \\ + 37 \\ \hline \end{array}
$$

10)
$$
\begin{array}{r} 9 \\ + 97 \\ \hline \end{array}
$$

11)
$$
\begin{array}{r} 64 \\ + 48 \\ \hline \end{array}
$$

12)
$$
\begin{array}{r} 5 \\ + 60 \\ \hline \end{array}
$$

13)
$$
\begin{array}{r} 66 \\ + 3 \\ \hline \end{array}
$$

14)
$$
\begin{array}{r} 20 \\ + 9 \\ \hline \end{array}
$$

15)
$$
\begin{array}{r} 95 \\ + 36 \\ \hline \end{array}
$$

16)
$$
\begin{array}{r} 77 \\ + 85 \\ \hline \end{array}
$$

17)
$$
\begin{array}{r} 59 \\ + 36 \\ \hline \end{array}
$$

18)
$$
\begin{array}{r} 31 \\ + 49 \\ \hline \end{array}
$$

19)
$$
\begin{array}{r} 37 \\ + 56 \\ \hline \end{array}
$$

20)
$$
\begin{array}{r} 95 \\ + 29 \\ \hline \end{array}
$$

Addition aplenty! Let's tackle these equations head-on!

1)
```
   35
+  69
_____
```

2)
```
   53
+   4
_____
```

3)
```
   46
+  46
_____
```

4)
```
   89
+  76
_____
```

5)
```
   46
+  40
_____
```

6)
```
   89
+  76
_____
```

7)
```
   80
+  23
_____
```

8)
```
   10
+  65
_____
```

9)
```
    1
+  74
_____
```

10)
```
   75
+  71
_____
```

11)
```
   20
+  84
_____
```

12)
```
   15
+  52
_____
```

13)
```
   31
+  59
_____
```

14)
```
   28
+  38
_____
```

15)
```
   22
+  67
_____
```

16)
```
   80
+  53
_____
```

17)
```
   97
+   2
_____
```

18)
```
    6
+  63
_____
```

19)
```
   20
+  25
_____
```

20)
```
   32
+   6
_____
```

 Adding up numbers like a pro! Let's sharpen those skills!

 **Day 10**

 **Name:** _____

**Date:** _____

 **Time Taken:** ___:___

**Score:** / **20**

1)
```
    7
+  84
―――
```

2)
```
   52
+  91
―――
```

3)
```
   63
+  73
―――
```

4)
```
   28
+  59
―――
```

5)
```
   61
+  31
―――
```

6)
```
   82
+   3
―――
```

7)
```
   29
+  69
―――
```

8)
```
   77
+  16
―――
```

9)
```
   23
+  54
―――
```

10)
```
   63
+  80
―――
```

11)
```
   26
+  87
―――
```

12)
```
   47
+  76
―――
```

13)
```
   86
+  21
―――
```

14)
```
   68
+  12
―――
```

15)
```
   65
+  31
―――
```

16)
```
    2
+  34
―――
```

17)
```
   67
+  50
―――
```

18)
```
   47
+  91
―――
```

19)
```
   33
+  92
―――
```

20)
```
   45
+   5
―――
```

 Mental math mode: engaged! Let's add quickly and accurately!

**Day 11**

 **Name:** _____

 **Time Taken:** ___:___

 **Date:** _____

**Score:** / 20

1) 87 + 19 =

2) 63 + 24 =

3) 19 + 88 =

4) 15 + 23 =

5) 84 + 82 =

6) 84 + 17 =

7) 23 + 33 =

8) 68 + 89 =

9) 27 + 28 =

10) 56 + 15 =

11) 45 + 57 =

12) 82 + 47 =

13) 82 + 96 =

14) 17 + 56 =

15) 16 + 20 =

16) 81 + 58 =

17) 42 + 67 =

18) 91 + 86 =

19) 87 + 31 =

20) 99 + 18 =

Ready to mentally add our way through today's challenges?

  **Name:** _____

 **Date:** _____

**Day 12**

 **Time Taken:** ___:___

 **Score:** / **20**

1) 31 + 36 =

2) 16 + 57 =

3) 75 + 30 =

4) 88 + 31 =

5) 86 + 54 =

6) 56 + 11 =

7) 92 + 79 =

8) 85 + 81 =

9) 29 + 43 =

10) 88 + 75 =

11) 79 + 76 =

12) 91 + 90 =

13) 93 + 91 =

14) 64 + 26 =

15) 68 + 36 =

16) 91 + 49 =

17) 51 + 13 =

18) 88 + 40 =

19) 68 + 99 =

20) 21 + 74 =

15

Let's sharpen our mental math skills with some quick addition!

1) $17 + 13 =$

2) $10 + 70 =$

3) $81 + 62 =$

4) $52 + 30 =$

5) $62 + 10 =$

6) $80 + 81 =$

7) $55 + 17 =$

8) $86 + 21 =$

9) $45 + 53 =$

10) $24 + 37 =$

11) $67 + 68 =$

12) $81 + 30 =$

13) $10 + 49 =$

14) $36 + 41 =$

15) $62 + 86 =$

16) $68 + 16 =$

17) $60 + 80 =$

18) $18 + 25 =$

19) $72 + 40 =$

20) $76 + 56 =$

Mental math magic time! Let's add with speed and accuracy!

**Day 14**

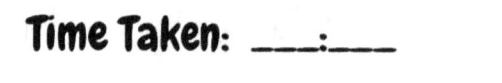

1)  70 + 89 =

2)  80 + 11 =

3)  14 + 79 =

4)  62 + 67 =

5)  76 + 31 =

6)  89 + 35 =

7)  15 + 25 =

8)  95 + 77 =

9)  97 + 30 =

10) 64 + 75 =

11) 79 + 32 =

12) 99 + 82 =

13) 70 + 46 =

14) 14 + 37 =

15) 97 + 36 =

16) 59 + 20 =

17) 91 + 53 =

18) 79 + 84 =

19) 64 + 55 =

20) 64 + 80 =

  **Name:** _____

 **Date:** _____

**Day 15**

 **Time Taken:** ___:___

 **Score:** / **20**

1) 21 + 76 =

2) 87 + 13 =

3) 48 + 56 =

4) 15 + 50 =

5) 17 + 64 =

6) 27 + 73 =

7) 10 + 35 =

8) 63 + 96 =

9) 16 + 63 =

10) 19 + 27 =

11) 95 + 56 =

12) 46 + 75 =

13) 11 + 90 =

14) 67 + 85 =

15) 48 + 18 =

16) 45 + 43 =

17) 62 + 24 =

18) 44 + 23 =

19) 60 + 60 =

20) 10 + 81 =

 Time to flex our mental muscles with some addition challenges!

  **Name:** _____

**Date:** _____

 **Time Taken:** ____:____

**Score:** / **20**

**Day 16**

1) 69 + 77 =

2) 11 + 42 =

3) 45 + 13 =

4) 15 + 44 =

5) 12 + 92 =

6) 81 + 22 =

7) 89 + 59 =

8) 29 + 55 =

9) 51 + 42 =

10) 15 + 66 =

11) 25 + 18 =

12) 42 + 29 =

13) 15 + 31 =

14) 90 + 63 =

15) 63 + 64 =

16) 92 + 94 =

17) 36 + 23 =

18) 30 + 83 =

19) 82 + 93 =

20) 78 + 10 =

19

 **Name:** _____

**Time Taken:** ___:___

 **Date:** _____

 **Score:** / 20

**Day 17**

1) 11 + 13 =

2) 51 + 78 =

3) 31 + 27 =

4) 85 + 14 =

5) 57 + 15 =

6) 15 + 71 =

7) 63 + 27 =

8) 20 + 88 =

9) 26 + 95 =

10) 13 + 64 =

11) 35 + 80 =

12) 84 + 71 =

13) 42 + 71 =

14) 92 + 40 =

15) 72 + 19 =

16) 28 + 31 =

17) 35 + 88 =

18) 65 + 52 =

19) 13 + 35 =

20) 39 + 70 =

Let's activate our mental math powers. Addition mode: on!

1) 25 + 15 =

2) 67 + 32 =

3) 42 + 98 =

4) 54 + 55 =

5) 72 + 40 =

6) 84 + 53 =

7) 27 + 80 =

8) 98 + 46 =

9) 32 + 84 =

10) 66 + 66 =

11) 99 + 59 =

12) 67 + 30 =

13) 11 + 56 =

14) 47 + 31 =

15) 92 + 60 =

16) 76 + 86 =

17) 91 + 87 =

18) 29 + 35 =

19) 63 + 26 =

20) 43 + 83 =

**Day 19**

Name: _____

Date: _____

Time Taken: ___:___

Score: ___ / 20

1) 21 + 60 =

2) 79 + 17 =

3) 75 + 37 =

4) 15 + 71 =

5) 58 + 72 =

6) 32 + 71 =

7) 93 + 16 =

8) 59 + 60 =

9) 26 + 85 =

10) 49 + 29 =

11) 91 + 75 =

12) 33 + 66 =

13) 42 + 95 =

14) 10 + 66 =

15) 37 + 84 =

16) 67 + 39 =

17) 27 + 59 =

18) 87 + 86 =

19) 15 + 72 =

20) 40 + 49 =

Ready to mentally add up some success? Let's do it!

1) 96 + 35 =

2) 39 + 55 =

3) 49 + 11 =

4) 28 + 85 =

5) 68 + 29 =

6) 47 + 52 =

7) 18 + 46 =

8) 46 + 38 =

9) 32 + 82 =

10) 60 + 91 =

11) 68 + 10 =

12) 56 + 13 =

13) 88 + 82 =

14) 76 + 85 =

15) 72 + 37 =

16) 30 + 71 =

17) 62 + 41 =

18) 68 + 84 =

19) 16 + 78 =

20) 89 + 47 =

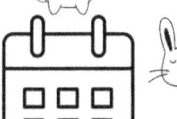

**Name:** _____

**Date:** _____

**Time Taken:** ___:___

**Score:** / 20

**Day 21**

1)
$$41 + \underline{\phantom{00}} = 127$$

2)
$$94 + \underline{\phantom{00}} = 173$$

3)
$$77 + \underline{\phantom{00}} = 120$$

4)
$$4 + \underline{\phantom{00}} = 10$$

5)
$$86 + \underline{\phantom{00}} = 131$$

6)
$$2 + \underline{\phantom{00}} = 86$$

7)
$$32 + \underline{\phantom{00}} = 61$$

8)
$$\phantom{0} + \underline{\phantom{00}} = 2$$

9)
$$63 + \underline{\phantom{00}} = 155$$

10)
$$33 + \underline{\phantom{00}} = 101$$

11)
$$70 + \underline{\phantom{00}} = 117$$

12)
$$3 + \underline{\phantom{00}} = 10$$

13)
$$75 + \underline{\phantom{00}} = 139$$

14)
$$1 + \underline{\phantom{00}} = 89$$

15)
$$84 + \underline{\phantom{00}} = 167$$

16)
$$6 + \underline{\phantom{00}} = 11$$

17)
$$95 + \underline{\phantom{00}} = 193$$

18)
$$84 + \underline{\phantom{00}} = 161$$

19)
$$88 + \underline{\phantom{00}} = 129$$

20)
$$9 + \underline{\phantom{00}} = 11$$

24

Time to play detective and find the missing numbers. Let's crack the code!

 **Name:** _____       **Date:** _____

**Day 22**  **Time Taken:** ____:____      **Score:** ____ / **20**

1)
```
   61
+  __
   88
```

2)
```
   92
+  __
  182
```

3)
```
   53
+  __
   85
```

4)
```
    4
+  __
   12
```

5)
```
   51
+  __
   77
```

6)
```
   53
+  __
   95
```

7)
```
    7
+  __
   98
```

8)
```
    2
+  __
    4
```

9)
```
   43
+  __
  130
```

10)
```
   16
+  __
   65
```

11)
```
   47
+  __
  143
```

12)
```
    6
+  __
   14
```

13)
```
   81
+  __
   92
```

14)
```
   23
+  __
   33
```

15)
```
   10
+  __
   21
```

16)
```
    3
+  __
    7
```

17)
```
   77
+  __
   84
```

18)
```
   35
+  __
   86
```

19)
```
   52
+  __
  147
```

20)
```
    3
+  __
    7
```

25

Missing numbers adding some mystery? Let's uncover the secrets!

  **Name:** _____    **Date:** _____

**Day 23**     **Time Taken:** ___:___    **Score:**    /    **20**

1)
```
   17
+ ___
  112
```

2)
```
    4
+ ___
  103
```

3)
```
   58
+ ___
  125
```

4)
```
    2
+ ___
   12
```

5)
```
   54
+ ___
   96
```

6)
```
   75
+ ___
   95
```

7)
```
   43
+ ___
   70
```

8)
```
    8
+ ___
   18
```

9)
```
   37
+ ___
   65
```

10)
```
   35
+ ___
   36
```

11)
```
   11
+ ___
   40
```

12)
```
    5
+ ___
   12
```

13)
```
    6
+ ___
   93
```

14)
```
   98
+ ___
  197
```

15)
```
   31
+ ___
  100
```

16)
```
    3
+ ___
    9
```

17)
```
   16
+ ___
   32
```

18)
```
   62
+ ___
  105
```

19)
```
   37
+ ___
  121
```

20)
```
    1
+ ___
    7
```

26

Let's complete the addition puzzle by filling in the missing pieces!

1)
$$\begin{array}{r} 79 \\ + \phantom{00} \\ \hline 99 \end{array}$$

2)
$$\begin{array}{r} 53 \\ + \phantom{00} \\ \hline 57 \end{array}$$

3)
$$\begin{array}{r} 87 \\ + \phantom{00} \\ \hline 137 \end{array}$$

4)
$$\begin{array}{r} \phantom{00} \\ + \phantom{00} \\ \hline 8 \end{array}$$

5)
$$\begin{array}{r} 75 \\ + \phantom{00} \\ \hline 148 \end{array}$$

6)
$$\begin{array}{r} 65 \\ + \phantom{00} \\ \hline 145 \end{array}$$

7)
$$\begin{array}{r} 38 \\ + \phantom{00} \\ \hline 91 \end{array}$$

8)
$$\begin{array}{r} 2 \\ + \phantom{00} \\ \hline 10 \end{array}$$

9)
$$\begin{array}{r} 36 \\ + \phantom{00} \\ \hline 49 \end{array}$$

10)
$$\begin{array}{r} 97 \\ + \phantom{00} \\ \hline 110 \end{array}$$

11)
$$\begin{array}{r} 12 \\ + \phantom{00} \\ \hline 60 \end{array}$$

12)
$$\begin{array}{r} 5 \\ + \phantom{00} \\ \hline 5 \end{array}$$

13)
$$\begin{array}{r} 29 \\ + \phantom{00} \\ \hline 125 \end{array}$$

14)
$$\begin{array}{r} 94 \\ + \phantom{00} \\ \hline 96 \end{array}$$

15)
$$\begin{array}{r} 98 \\ + \phantom{00} \\ \hline 110 \end{array}$$

16)
$$\begin{array}{r} 8 \\ + \phantom{00} \\ \hline 16 \end{array}$$

17)
$$\begin{array}{r} 41 \\ + \phantom{00} \\ \hline 87 \end{array}$$

18)
$$\begin{array}{r} 21 \\ + \phantom{00} \\ \hline 96 \end{array}$$

19)
$$\begin{array}{r} 24 \\ + \phantom{00} \\ \hline 106 \end{array}$$

20)
$$\begin{array}{r} 8 \\ + \phantom{00} \\ \hline 14 \end{array}$$

Adding numbers is fun, especially when there's a mystery involved!

**Day 25**

 **Name:** _____

**Date:** _____

 **Time Taken:** ___:___

**Score:** / 20

1)
$$\begin{array}{r} 27 \\ +\phantom{00} \\ \hline 77 \end{array}$$

2)
$$\begin{array}{r} 89 \\ +\phantom{00} \\ \hline 106 \end{array}$$

3)
$$\begin{array}{r} 1 \\ +\phantom{00} \\ \hline 81 \end{array}$$

4)
$$\begin{array}{r} 8 \\ +\phantom{00} \\ \hline 17 \end{array}$$

5)
$$\begin{array}{r} 6 \\ +\phantom{00} \\ \hline 81 \end{array}$$

6)
$$\begin{array}{r} 18 \\ +\phantom{00} \\ \hline 39 \end{array}$$

7)
$$\begin{array}{r} 98 \\ +\phantom{00} \\ \hline 163 \end{array}$$

8)
$$\begin{array}{r} 2 \\ +\phantom{00} \\ \hline 11 \end{array}$$

9)
$$\begin{array}{r} 36 \\ +\phantom{00} \\ \hline 50 \end{array}$$

10)
$$\begin{array}{r} 57 \\ +\phantom{00} \\ \hline 110 \end{array}$$

11)
$$\begin{array}{r} 62 \\ +\phantom{00} \\ \hline 84 \end{array}$$

12)
$$\begin{array}{r} 4 \\ +\phantom{00} \\ \hline 6 \end{array}$$

13)
$$\begin{array}{r} 12 \\ +\phantom{00} \\ \hline 110 \end{array}$$

14)
$$\begin{array}{r} 93 \\ +\phantom{00} \\ \hline 186 \end{array}$$

15)
$$\begin{array}{r} 80 \\ +\phantom{00} \\ \hline 119 \end{array}$$

16)
$$\begin{array}{r} 8 \\ +\phantom{00} \\ \hline 16 \end{array}$$

17)
$$\begin{array}{r} 87 \\ +\phantom{00} \\ \hline 181 \end{array}$$

18)
$$\begin{array}{r} 21 \\ +\phantom{00} \\ \hline 66 \end{array}$$

19)
$$\begin{array}{r} 25 \\ +\phantom{00} \\ \hline 41 \end{array}$$

20)
$$\begin{array}{r} 3 \\ +\phantom{00} \\ \hline 6 \end{array}$$

Ready to put our detective hats on and find the missing numbers?

 **Name:** _____

 **Time Taken:** ___:___

**Date:** _____

**Score:** / **20**

1)
$$74 + \underline{\phantom{00}} \\ 104$$

2)
$$60 + \underline{\phantom{00}} \\ 88$$

3)
$$40 + \underline{\phantom{00}} \\ 130$$

4)
$$6 + \underline{\phantom{00}} \\ 14$$

5)
$$30 + \underline{\phantom{00}} \\ 112$$

6)
$$84 + \underline{\phantom{00}} \\ 144$$

7)
$$55 + \underline{\phantom{00}} \\ 148$$

8)
$$2 + \underline{\phantom{00}} \\ 2$$

9)
$$4 + \underline{\phantom{00}} \\ 41$$

10)
$$22 + \underline{\phantom{00}} \\ 87$$

11)
$$75 + \underline{\phantom{00}} \\ 166$$

12)
$$8 + \underline{\phantom{00}} \\ 16$$

13)
$$89 + \underline{\phantom{00}} \\ 134$$

14)
$$9 + \underline{\phantom{00}} \\ 105$$

15)
$$61 + \underline{\phantom{00}} \\ 136$$

16)
$$8 + \underline{\phantom{00}} \\ 10$$

17)
$$91 + \underline{\phantom{00}} \\ 126$$

18)
$$57 + \underline{\phantom{00}} \\ 84$$

19)
$$17 + \underline{\phantom{00}} \\ 67$$

20)
$$3 + \underline{\phantom{00}} \\ 4$$

**Name:** _____

**Date:** _____

**Day 27**

**Time Taken:** ___:___

**Score:** ___ / **20**

1)
$$37 + \underline{\phantom{00}} \over 103$$

2)
$$30 + \underline{\phantom{00}} \over 58$$

3)
$$53 + \underline{\phantom{00}} \over 147$$

4)
$$1 + \underline{\phantom{00}} \over 3$$

5)
$$79 + \underline{\phantom{00}} \over 153$$

6)
$$66 + \underline{\phantom{00}} \over 70$$

7)
$$42 + \underline{\phantom{00}} \over 93$$

8)
$$6 + \underline{\phantom{00}} \over 11$$

9)
$$45 + \underline{\phantom{00}} \over 129$$

10)
$$40 + \underline{\phantom{00}} \over 122$$

11)
$$59 + \underline{\phantom{00}} \over 116$$

12)
$$4 + \underline{\phantom{00}} \over 5$$

13)
$$56 + \underline{\phantom{00}} \over 83$$

14)
$$85 + \underline{\phantom{00}} \over 172$$

15)
$$14 + \underline{\phantom{00}} \over 56$$

16)
$$9 + \underline{\phantom{00}} \over 18$$

17)
$$89 + \underline{\phantom{00}} \over 94$$

18)
$$38 + \underline{\phantom{00}} \over 67$$

19)
$$92 + \underline{\phantom{00}} \over 138$$

20)
$$6 + \underline{\phantom{00}} \over 7$$

Time to solve some addition mysteries. Let's find those missing pieces!

 **Name:** _____

 **Date:** _____

**Day 28**

**Time Taken:** ___:___

 **Score:** / **20**

1)
```
    10
  +
  ―――
    87
```

2)
```
    65
  +
  ―――
    76
```

3)
```
    23
  +
  ―――
    45
```

4)
```
     9
  +
  ―――
    14
```

5)
```
    66
  +
  ―――
   111
```

6)
```
    97
  +
  ―――
   170
```

7)
```
    18
  +
  ―――
    83
```

8)
```
     5
  +
  ―――
    15
```

9)
```
    73
  +
  ―――
   103
```

10)
```
    76
  +
  ―――
   129
```

11)
```
    30
  +
  ―――
   116
```

12)
```
     5
  +
  ―――
    12
```

13)
```
    79
  +
  ―――
   144
```

14)
```
    74
  +
  ―――
   136
```

15)
```
    60
  +
  ―――
   157
```

16)
```
     8
  +
  ―――
    16
```

17)
```
    65
  +
  ―――
   156
```

18)
```
    13
  +
  ―――
   104
```

19)
```
    63
  +
  ―――
   153
```

20)
```
     5
  +
  ―――
     6
```

**Day 29**

 **Name:** _____

 **Time Taken:** ___:___

 **Date:** _____

**Score:** ___ / **20**

1)
$$\begin{array}{r} 92 \\ + \phantom{00} \\ \hline 163 \end{array}$$

2)
$$\begin{array}{r} 8 \\ + \phantom{00} \\ \hline 58 \end{array}$$

3)
$$\begin{array}{r} 40 \\ + \phantom{00} \\ \hline 88 \end{array}$$

4)
$$\begin{array}{r} 5 \\ + \phantom{00} \\ \hline 8 \end{array}$$

5)
$$\begin{array}{r} 28 \\ + \phantom{00} \\ \hline 93 \end{array}$$

6)
$$\begin{array}{r} 74 \\ + \phantom{00} \\ \hline 171 \end{array}$$

7)
$$\begin{array}{r} 13 \\ + \phantom{00} \\ \hline 74 \end{array}$$

8)
$$\begin{array}{r} 7 \\ + \phantom{00} \\ \hline 15 \end{array}$$

9)
$$\begin{array}{r} 54 \\ + \phantom{00} \\ \hline 73 \end{array}$$

10)
$$\begin{array}{r} 32 \\ + \phantom{00} \\ \hline 65 \end{array}$$

11)
$$\begin{array}{r} 75 \\ + \phantom{00} \\ \hline 100 \end{array}$$

12)
$$\begin{array}{r} 8 \\ + \phantom{00} \\ \hline 17 \end{array}$$

13)
$$\begin{array}{r} 64 \\ + \phantom{00} \\ \hline 109 \end{array}$$

14)
$$\begin{array}{r} 37 \\ + \phantom{00} \\ \hline 86 \end{array}$$

15)
$$\begin{array}{r} 42 \\ + \phantom{00} \\ \hline 69 \end{array}$$

16)
$$\begin{array}{r} 4 \\ + \phantom{00} \\ \hline 8 \end{array}$$

17)
$$\begin{array}{r} 99 \\ + \phantom{00} \\ \hline 133 \end{array}$$

18)
$$\begin{array}{r} 23 \\ + \phantom{00} \\ \hline 62 \end{array}$$

19)
$$\begin{array}{r} 69 \\ + \phantom{00} \\ \hline 162 \end{array}$$

20)
$$\begin{array}{r} 1 \\ + \phantom{00} \\ \hline 5 \end{array}$$

Let's add up the excitement by uncovering the missing numbers!

**Day 30**

Name: _____

Date: _____

Time Taken: ___:___

Score: ___ / 20

1)
```
   34
 +
   56
```

2)
```
   70
 +
  100
```

3)
```
   54
 +
  140
```

4)
```
    5
 +
    9
```

5)
```
   50
 +
  105
```

6)
```
   12
 +
   84
```

7)
```
   99
 +
  135
```

8)
```
    4
 +
    6
```

9)
```
   54
 +
   95
```

10)
```
   16
 +
   35
```

11)
```
   69
 +
   84
```

12)
```
    8
 +
    8
```

13)
```
   59
 +
   62
```

14)
```
   28
 +
   42
```

15)
```
    9
 +
   90
```

16)
```
    3
 +
   11
```

17)
```
   33
 +
   66
```

18)
```
   78
 +
  156
```

19)
```
   46
 +
   65
```

20)
```
    4
 +
    4
```

Time to subtract and see what's left. Let's get started!

**Day 31**

 **Name:** _____     **Date:** _____

 **Time Taken:** ___:___     **Score:**    /  **20**

1)
$$\begin{array}{r} 51 \\ -\ 7 \\ \hline \end{array}$$

2)
$$\begin{array}{r} 68 \\ -\ 64 \\ \hline \end{array}$$

3)
$$\begin{array}{r} 65 \\ -\ 38 \\ \hline \end{array}$$

4)
$$\begin{array}{r} 9 \\ -\ 1 \\ \hline \end{array}$$

5)
$$\begin{array}{r} 44 \\ -\ 25 \\ \hline \end{array}$$

6)
$$\begin{array}{r} 59 \\ -\ 40 \\ \hline \end{array}$$

7)
$$\begin{array}{r} 76 \\ -\ 35 \\ \hline \end{array}$$

8)
$$\begin{array}{r} 7 \\ -\ 5 \\ \hline \end{array}$$

9)
$$\begin{array}{r} 83 \\ -\ 65 \\ \hline \end{array}$$

10)
$$\begin{array}{r} 62 \\ -\ 61 \\ \hline \end{array}$$

11)
$$\begin{array}{r} 62 \\ -\ 36 \\ \hline \end{array}$$

12)
$$\begin{array}{r} 7 \\ -\ 5 \\ \hline \end{array}$$

13)
$$\begin{array}{r} 45 \\ -\ 11 \\ \hline \end{array}$$

14)
$$\begin{array}{r} 97 \\ -\ 74 \\ \hline \end{array}$$

15)
$$\begin{array}{r} 51 \\ -\ 28 \\ \hline \end{array}$$

16)
$$\begin{array}{r} 8 \\ -\ 3 \\ \hline \end{array}$$

17)
$$\begin{array}{r} 88 \\ -\ 52 \\ \hline \end{array}$$

18)
$$\begin{array}{r} 49 \\ -\ 24 \\ \hline \end{array}$$

19)
$$\begin{array}{r} 32 \\ -\ 24 \\ \hline \end{array}$$

20)
$$\begin{array}{r} 8 \\ -\ 4 \\ \hline \end{array}$$

Let's simplify things by subtracting the unnecessary. Ready?

1)
```
   95
-  58
_____
```

2)
```
   49
-  12
_____
```

3)
```
   77
-  45
_____
```

4)
```
    7
-
_____
```

5)
```
   64
-   9
_____
```

6)
```
   44
-  44
_____
```

7)
```
   59
-  41
_____
```

8)
```
    8
-   6
_____
```

9)
```
   90
-  81
_____
```

10)
```
   86
-  71
_____
```

11)
```
   82
-  51
_____
```

12)
```
    7
-   2
_____
```

13)
```
   75
-  73
_____
```

14)
```
   73
-  36
_____
```

15)
```
   84
-  22
_____
```

16)
```
    2
-
_____
```

17)
```
   85
-  79
_____
```

18)
```
   93
-  10
_____
```

19)
```
   52
-  26
_____
```

20)
```
    4
-   1
_____
```

Subtraction time! Let's subtract and simplify our equations!

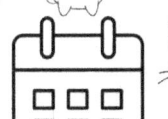

1)
$$\begin{array}{r} 60 \\ -\ 34 \\ \hline \end{array}$$

2)
$$\begin{array}{r} 83 \\ -\ 81 \\ \hline \end{array}$$

3)
$$\begin{array}{r} 68 \\ -\ 16 \\ \hline \end{array}$$

4)
$$\begin{array}{r} 8 \\ -\ 8 \\ \hline \end{array}$$

5)
$$\begin{array}{r} 77 \\ -\ 14 \\ \hline \end{array}$$

6)
$$\begin{array}{r} 58 \\ -\ 44 \\ \hline \end{array}$$

7)
$$\begin{array}{r} 21 \\ -\ 13 \\ \hline \end{array}$$

8)
$$\begin{array}{r} 9 \\ -\ 7 \\ \hline \end{array}$$

9)
$$\begin{array}{r} 52 \\ -\ 36 \\ \hline \end{array}$$

10)
$$\begin{array}{r} 76 \\ -\ 41 \\ \hline \end{array}$$

11)
$$\begin{array}{r} 54 \\ -\ 50 \\ \hline \end{array}$$

12)
$$\begin{array}{r} 8 \\ -\ 2 \\ \hline \end{array}$$

13)
$$\begin{array}{r} 71 \\ -\ 39 \\ \hline \end{array}$$

14)
$$\begin{array}{r} 76 \\ -\ 71 \\ \hline \end{array}$$

15)
$$\begin{array}{r} 37 \\ -\ 34 \\ \hline \end{array}$$

16)
$$\begin{array}{r} 9 \\ -\ 8 \\ \hline \end{array}$$

17)
$$\begin{array}{r} 45 \\ -\ 7 \\ \hline \end{array}$$

18)
$$\begin{array}{r} 79 \\ -\ 69 \\ \hline \end{array}$$

19)
$$\begin{array}{r} 69 \\ -\ 34 \\ \hline \end{array}$$

20)
$$\begin{array}{r} 1 \\ -\ \\ \hline \end{array}$$

Let's subtract and streamline our way to success!

**Day 34**

 **Name:** _____

**Time Taken:** ___:___

 **Date:** _____

 **Score:**        /   **20**

1)
$$\begin{array}{r} 38 \\ -\ 36 \\ \hline \end{array}$$

2)
$$\begin{array}{r} 74 \\ -\ 33 \\ \hline \end{array}$$

3)
$$\begin{array}{r} 50 \\ -\ \ 8 \\ \hline \end{array}$$

4)
$$\begin{array}{r} 5 \\ -\ 1 \\ \hline \end{array}$$

5)
$$\begin{array}{r} 86 \\ -\ 56 \\ \hline \end{array}$$

6)
$$\begin{array}{r} 67 \\ -\ 41 \\ \hline \end{array}$$

7)
$$\begin{array}{r} 98 \\ -\ 12 \\ \hline \end{array}$$

8)
$$\begin{array}{r} 5 \\ -\ \\ \hline \end{array}$$

9)
$$\begin{array}{r} 69 \\ -\ 56 \\ \hline \end{array}$$

10)
$$\begin{array}{r} 99 \\ -\ 33 \\ \hline \end{array}$$

11)
$$\begin{array}{r} 84 \\ -\ 46 \\ \hline \end{array}$$

12)
$$\begin{array}{r} 3 \\ -\ 3 \\ \hline \end{array}$$

13)
$$\begin{array}{r} 51 \\ -\ 28 \\ \hline \end{array}$$

14)
$$\begin{array}{r} 82 \\ -\ 50 \\ \hline \end{array}$$

15)
$$\begin{array}{r} 20 \\ -\ 17 \\ \hline \end{array}$$

16)
$$\begin{array}{r} 2 \\ -\ 1 \\ \hline \end{array}$$

17)
$$\begin{array}{r} 49 \\ -\ 28 \\ \hline \end{array}$$

18)
$$\begin{array}{r} 82 \\ -\ 15 \\ \hline \end{array}$$

19)
$$\begin{array}{r} 60 \\ -\ \ 7 \\ \hline \end{array}$$

20)
$$\begin{array}{r} 6 \\ -\ \\ \hline \end{array}$$

Subtracting numbers is like peeling layers. Let's peel away!

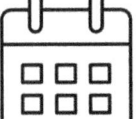

1)
$$\begin{array}{r} 68 \\ -\ 60 \\ \hline \end{array}$$

2)
$$\begin{array}{r} 73 \\ -\ 60 \\ \hline \end{array}$$

3)
$$\begin{array}{r} 97 \\ -\ 30 \\ \hline \end{array}$$

4)
$$\begin{array}{r} 9 \\ -\ 1 \\ \hline \end{array}$$

5)
$$\begin{array}{r} 35 \\ -\ 10 \\ \hline \end{array}$$

6)
$$\begin{array}{r} 69 \\ -\ 66 \\ \hline \end{array}$$

7)
$$\begin{array}{r} 43 \\ -\ 27 \\ \hline \end{array}$$

8)
$$\begin{array}{r} 8 \\ -\ 6 \\ \hline \end{array}$$

9)
$$\begin{array}{r} 50 \\ -\ 19 \\ \hline \end{array}$$

10)
$$\begin{array}{r} 81 \\ -\ 22 \\ \hline \end{array}$$

11)
$$\begin{array}{r} 87 \\ -\ 70 \\ \hline \end{array}$$

12)
$$\begin{array}{r} 8 \\ -\ 1 \\ \hline \end{array}$$

13)
$$\begin{array}{r} 91 \\ -\ 56 \\ \hline \end{array}$$

14)
$$\begin{array}{r} 87 \\ -\ 24 \\ \hline \end{array}$$

15)
$$\begin{array}{r} 89 \\ -\ 17 \\ \hline \end{array}$$

16)
$$\begin{array}{r} 9 \\ -\ 6 \\ \hline \end{array}$$

17)
$$\begin{array}{r} 86 \\ -\ 31 \\ \hline \end{array}$$

18)
$$\begin{array}{r} 78 \\ -\ 14 \\ \hline \end{array}$$

19)
$$\begin{array}{r} 65 \\ -\ 39 \\ \hline \end{array}$$

20)
$$\begin{array}{r} 3 \\ -\ 3 \\ \hline \end{array}$$

Let's subtract and declutter our math problems!

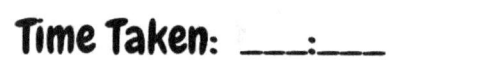
1)
$$\begin{array}{r} 62 \\ -\ \ 9 \\ \hline \end{array}$$

2)
$$\begin{array}{r} 35 \\ -\ 27 \\ \hline \end{array}$$

3)
$$\begin{array}{r} 81 \\ -\ 37 \\ \hline \end{array}$$

4)
$$\begin{array}{r} 3 \\ -\ 1 \\ \hline \end{array}$$

5)
$$\begin{array}{r} 99 \\ -\ 65 \\ \hline \end{array}$$

6)
$$\begin{array}{r} 12 \\ -\ \ 6 \\ \hline \end{array}$$

7)
$$\begin{array}{r} 79 \\ -\ 48 \\ \hline \end{array}$$

8)
$$\begin{array}{r} 4 \\ -\ 3 \\ \hline \end{array}$$

9)
$$\begin{array}{r} 72 \\ -\ 52 \\ \hline \end{array}$$

10)
$$\begin{array}{r} 61 \\ -\ 14 \\ \hline \end{array}$$

11)
$$\begin{array}{r} 57 \\ -\ 41 \\ \hline \end{array}$$

12)
$$\begin{array}{r} 7 \\ -\ 4 \\ \hline \end{array}$$

13)
$$\begin{array}{r} 73 \\ -\ 46 \\ \hline \end{array}$$

14)
$$\begin{array}{r} 92 \\ -\ 16 \\ \hline \end{array}$$

15)
$$\begin{array}{r} 24 \\ -\ \ 1 \\ \hline \end{array}$$

16)
$$\begin{array}{r} 6 \\ -\ 4 \\ \hline \end{array}$$

17)
$$\begin{array}{r} 51 \\ -\ 24 \\ \hline \end{array}$$

18)
$$\begin{array}{r} 55 \\ -\ 30 \\ \hline \end{array}$$

19)
$$\begin{array}{r} 66 \\ -\ 19 \\ \hline \end{array}$$

20)
$$\begin{array}{r} 9 \\ -\ 4 \\ \hline \end{array}$$

Subtraction is about finding balance. Let's find our equilibrium!

 **Name:** _____

 **Date:** _____

 **Time Taken:** ____:____

 **Score:** _____ / **20**

1)
```
   96
-  82
-----
```

2)
```
   74
-   7
-----
```

3)
```
   80
-  28
-----
```

4)
```
    7
-   3
-----
```

5)
```
   75
-  44
-----
```

6)
```
   35
-   4
-----
```

7)
```
   96
-  34
-----
```

8)
```
    8
-   6
-----
```

9)
```
   69
-  54
-----
```

10)
```
   82
-   2
-----
```

11)
```
   80
-  17
-----
```

12)
```
    9
-   7
-----
```

13)
```
   42
-  35
-----
```

14)
```
   52
-  40
-----
```

15)
```
   44
-  40
-----
```

16)
```
    4
-   2
-----
```

17)
```
   63
-  54
-----
```

18)
```
   55
-  14
-----
```

19)
```
   77
-  50
-----
```

20)
```
    9
-   6
-----
```

 **Name:** _____

**Time Taken:** ___:___

 **Date:** _____

 **Score:** / 20

**Day 38**

1)
```
   43
 -  6
_____
```

2)
```
   87
 - 79
_____
```

3)
```
   59
 - 23
_____
```

4)
```
    7
 -  4
_____
```

5)
```
   72
 -  5
_____
```

6)
```
   67
 - 58
_____
```

7)
```
   49
 - 21
_____
```

8)
```
    3
 -
_____
```

9)
```
   38
 - 35
_____
```

10)
```
   83
 - 63
_____
```

11)
```
   62
 - 15
_____
```

12)
```
    9
 -  5
_____
```

13)
```
   34
 - 10
_____
```

14)
```
   62
 - 11
_____
```

15)
```
   87
 - 53
_____
```

16)
```
    5
 -  5
_____
```

17)
```
   14
 -  3
_____
```

18)
```
   64
 -  9
_____
```

19)
```
   74
 - 13
_____
```

20)
```
    1
 -  1
_____
```

  **Name:** _____     **Date:** _____

**Day 39**   **Time Taken:** ___:___     **Score:**    / 20

1)
```
   46
 - 25
```

2)
```
   55
 -  4
```

3)
```
   60
 - 14
```

4)
```
    9
 -  9
```

5)
```
   74
 - 30
```

6)
```
   76
 - 36
```

7)
```
   65
 -  6
```

8)
```
    5
 -  1
```

9)
```
   86
 - 64
```

10)
```
   92
 - 56
```

11)
```
   69
 - 41
```

12)
```
    7
 -  2
```

13)
```
   59
 - 55
```

14)
```
   83
 - 68
```

15)
```
   99
 - 98
```

16)
```
    9
 -  3
```

17)
```
   99
 - 51
```

18)
```
   76
 - 45
```

19)
```
   94
 - 80
```

20)
```
    8
 -  3
```

Subtraction simplifies. Let's simplify our equations together!

Day 40

 Name: _____

 Time Taken: ___:___

 Date: _____

Score: ___ / 20

1)
$$\begin{array}{r} 99 \\ -\ 84 \\ \hline \end{array}$$

2)
$$\begin{array}{r} 27 \\ -\ 13 \\ \hline \end{array}$$

3)
$$\begin{array}{r} 95 \\ -\ 40 \\ \hline \end{array}$$

4)
$$\begin{array}{r} 9 \\ -\ 6 \\ \hline \end{array}$$

5)
$$\begin{array}{r} 44 \\ -\ 5 \\ \hline \end{array}$$

6)
$$\begin{array}{r} 68 \\ -\ 26 \\ \hline \end{array}$$

7)
$$\begin{array}{r} 93 \\ -\ 26 \\ \hline \end{array}$$

8)
$$\begin{array}{r} 3 \\ -\ 3 \\ \hline \end{array}$$

9)
$$\begin{array}{r} 95 \\ -\ 23 \\ \hline \end{array}$$

10)
$$\begin{array}{r} 62 \\ -\ 33 \\ \hline \end{array}$$

11)
$$\begin{array}{r} 88 \\ -\ 59 \\ \hline \end{array}$$

12)
$$\begin{array}{r} 7 \\ -\ 3 \\ \hline \end{array}$$

13)
$$\begin{array}{r} 84 \\ -\ 16 \\ \hline \end{array}$$

14)
$$\begin{array}{r} 51 \\ -\ 26 \\ \hline \end{array}$$

15)
$$\begin{array}{r} 64 \\ -\ 33 \\ \hline \end{array}$$

16)
$$\begin{array}{r} 4 \\ -\ 1 \\ \hline \end{array}$$

17)
$$\begin{array}{r} 10 \\ -\ 6 \\ \hline \end{array}$$

18)
$$\begin{array}{r} 81 \\ -\ 24 \\ \hline \end{array}$$

19)
$$\begin{array}{r} 81 \\ -\ 57 \\ \hline \end{array}$$

20)
$$\begin{array}{r} 9 \\ -\ 4 \\ \hline \end{array}$$

 **Name:** _____

**Time Taken:** ____:____

 **Date:** _____

 **Score:** ___ / 20

**Day 41**

1) 82 − 50 =

2) 60 − 32 =

3) 63 − 26 =

4) 71 − 10 =

5) 42 − 11 =

6) 66 − 25 =

7) 43 − 22 =

8) 40 − 38 =

9) 72 − 40 =

10) 81 − 47 =

11) 98 − 29 =

12) 75 − 29 =

13) 74 − 70 =

14) 85 − 22 =

15) 83 − 10 =

16) 80 − 13 =

17) 84 − 46 =

18) 55 − 39 =

19) 88 − 40 =

20) 95 − 59 =

**Name:** _____

**Date:** _____

**Day 42**

**Time Taken:** ___:___

**Score:** ___ / 20

1) 23 − 16 =

2) 64 − 13 =

3) 56 − 19 =

4) 91 − 66 =

5) 97 − 36 =

6) 53 − 50 =

7) 72 − 38 =

8) 99 − 19 =

9) 71 − 61 =

10) 88 − 86 =

11) 80 − 28 =

12) 26 − 17 =

13) 71 − 31 =

14) 79 − 30 =

15) 84 − 24 =

16) 62 − 29 =

17) 92 − 60 =

18) 95 − 79 =

19) 90 − 75 =

20) 75 − 24 =

45

**Name:** _____

**Date:** _____

**Day 43**

**Time Taken:** ___:___

**Score:** / 20

1) $86 - 62 =$

2) $95 - 64 =$

3) $30 - 21 =$

4) $88 - 33 =$

5) $72 - 65 =$

6) $47 - 43 =$

7) $88 - 58 =$

8) $38 - 37 =$

9) $93 - 67 =$

10) $98 - 92 =$

11) $48 - 11 =$

12) $99 - 98 =$

13) $86 - 45 =$

14) $47 - 45 =$

15) $89 - 22 =$

16) $87 - 59 =$

17) $83 - 58 =$

18) $86 - 65 =$

19) $57 - 24 =$

20) $80 - 55 =$

Ready to subtract quickly and accurately in our minds? Let's do it!

 **Name:** _____

 **Date:** _____

**Day 44** **Time Taken:** ___:___

 **Score:** / **20**

1) 96 – 53 =

2) 95 – 77 =

3) 58 – 39 =

4) 99 – 34 =

5) 57 – 56 =

6) 48 – 18 =

7) 90 – 50 =

8) 68 – 22 =

9) 55 – 48 =

10) 66 – 14 =

11) 81 – 74 =

12) 72 – 37 =

13) 63 – 39 =

14) 96 – 10 =

15) 90 – 43 =

16) 85 – 72 =

17) 46 – 42 =

18) 53 – 19 =

19) 37 – 21 =

20) 95 – 66 =

 **Name:** _____

 **Date:** _____

**Day 45**

 **Time Taken:** ____:____

 **Score:** ___ / 20

1) $82 - 63 =$

2) $63 - 20 =$

3) $44 - 20 =$

4) $96 - 44 =$

5) $83 - 10 =$

6) $82 - 69 =$

7) $86 - 48 =$

8) $86 - 74 =$

9) $72 - 37 =$

10) $61 - 35 =$

11) $89 - 45 =$

12) $51 - 19 =$

13) $93 - 37 =$

14) $79 - 66 =$

15) $92 - 48 =$

16) $64 - 59 =$

17) $82 - 33 =$

18) $76 - 43 =$

19) $41 - 21 =$

20) $98 - 41 =$

Time to mentally subtract and simplify. Let's sharpen those skills!

 **Name:** _____

 **Date:** _____

**Day 46**

**Time Taken:** ___:___

 **Score:** / **20**

1) 80 - 72 =

2) 60 - 18 =

3) 41 - 24 =

4) 74 - 39 =

5) 26 - 23 =

6) 25 - 14 =

7) 90 - 57 =

8) 52 - 51 =

9) 47 - 11 =

10) 93 - 70 =

11) 68 - 31 =

12) 37 - 12 =

13) 75 - 49 =

14) 92 - 68 =

15) 91 - 31 =

16) 95 - 90 =

17) 67 - 38 =

18) 99 - 20 =

19) 63 - 17 =

20) 93 - 28 =

Let's activate our mental math powers. Subtraction mode: on!

Name: _____

Date: _____

Time Taken: ___:___

Score: ___ / 20

1) 88 – 55 =

2) 32 – 30 =

3) 58 – 20 =

4) 51 – 13 =

5) 84 – 56 =

6) 44 – 39 =

7) 88 – 88 =

8) 88 – 64 =

9) 68 – 26 =

10) 85 – 58 =

11) 42 – 24 =

12) 54 – 45 =

13) 76 – 65 =

14) 58 – 48 =

15) 65 – 64 =

16) 48 – 32 =

17) 18 – 18 =

18) 15 – 13 =

19) 89 – 63 =

20) 42 – 12 =

 Mental math masters, unite! Let's tackle these subtraction problems!

  **Name:** _____     **Date:** _____

**Day 48**     **Time Taken:** ___:___     **Score:**    / 20

1) 40 − 25 =

2) 25 − 25 =

3) 85 − 82 =

4) 75 − 66 =

5) 55 − 50 =

6) 72 − 49 =

7) 80 − 56 =

8) 84 − 53 =

9) 59 − 54 =

10) 80 − 43 =

11) 49 − 39 =

12) 90 − 25 =

13) 74 − 27 =

14) 92 − 37 =

15) 73 − 25 =

16) 50 − 41 =

17) 65 − 47 =

18) 80 − 56 =

19) 72 − 32 =

20) 76 − 51 =

**Name:** _____

**Date:** _____

**Day 49**

**Time Taken:** ___:___

**Score:** ___ / 20

1) 92 − 15 =

2) 88 − 13 =

3) 94 − 31 =

4) 97 − 39 =

5) 67 − 48 =

6) 87 − 80 =

7) 93 − 79 =

8) 37 − 24 =

9) 53 − 24 =

10) 62 − 50 =

11) 80 − 67 =

12) 72 − 10 =

13) 65 − 48 =

14) 80 − 33 =

15) 77 − 29 =

16) 76 − 11 =

17) 97 − 19 =

18) 48 − 18 =

19) 55 − 33 =

20) 84 − 83 =

Mental subtraction mavens, assemble! Let's conquer these equations!

 **Name:** _____

 **Date:** _____

**Day 50**

 **Time Taken:** ___:___

 **Score:** / 20

1) 79 – 26 =

2) 88 – 52 =

3) 78 – 22 =

4) 92 – 36 =

5) 40 – 34 =

6) 55 – 27 =

7) 18 – 14 =

8) 83 – 68 =

9) 76 – 33 =

10) 66 – 19 =

11) 87 – 20 =

12) 54 – 38 =

13) 92 – 38 =

14) 42 – 33 =

15) 79 – 55 =

16) 70 – 38 =

17) 83 – 14 =

18) 34 – 34 =

19) 66 – 31 =

20) 70 – 43 =

Can you find the missing numbers? Let's crack the subtraction code!

1)
$$31$$
$$-$$
$$\overline{29}$$

2)
$$66$$
$$-$$
$$\overline{63}$$

3)
$$47$$
$$-$$
$$\overline{17}$$

4)
$$8$$
$$-$$
$$\overline{2}$$

5)
$$80$$
$$-$$
$$\overline{5}$$

6)
$$86$$
$$-$$
$$\overline{45}$$

7)
$$89$$
$$-$$
$$\overline{47}$$

8)
$$9$$
$$-$$
$$\overline{6}$$

9)
$$93$$
$$-$$
$$\overline{33}$$

10)
$$74$$
$$-$$
$$\overline{34}$$

11)
$$70$$
$$-$$
$$\overline{3}$$

12)
$$9$$
$$-$$
$$\overline{5}$$

13)
$$72$$
$$-$$
$$\overline{42}$$

14)
$$60$$
$$-$$
$$\overline{37}$$

15)
$$77$$
$$-$$
$$\overline{0}$$

16)
$$9$$
$$-$$
$$\overline{7}$$

17)
$$25$$
$$-$$
$$\overline{2}$$

18)
$$88$$
$$-$$
$$\overline{33}$$

19)
$$83$$
$$-$$
$$\overline{60}$$

20)
$$9$$
$$-$$
$$\overline{\phantom{0}}$$

54

  Time to play detective and uncover the missing numbers. Let's solve the mystery!

 **Name:** _____   **Date:** _____

 **Time Taken:** ___:___  **Score:**    /   **20**

**Day 52**

1)
$$\begin{array}{r} 52 \\ - \phantom{00} \\ \hline 21 \end{array}$$

2)
$$\begin{array}{r} 98 \\ - \phantom{00} \\ \hline 82 \end{array}$$

3)
$$\begin{array}{r} 50 \\ - \phantom{0} \\ \hline 8 \end{array}$$

4)
$$\begin{array}{r} 3 \\ - \phantom{0} \\ \hline 2 \end{array}$$

5)
$$\begin{array}{r} 92 \\ - \phantom{00} \\ \hline 54 \end{array}$$

6)
$$\begin{array}{r} 59 \\ - \phantom{00} \\ \hline 25 \end{array}$$

7)
$$\begin{array}{r} 79 \\ - \phantom{0} \\ \hline 1 \end{array}$$

8)
$$\begin{array}{r} 5 \\ - \phantom{0} \\ \hline 3 \end{array}$$

9)
$$\begin{array}{r} 77 \\ - \phantom{00} \\ \hline 68 \end{array}$$

10)
$$\begin{array}{r} 49 \\ - \phantom{00} \\ \hline 13 \end{array}$$

11)
$$\begin{array}{r} 95 \\ - \phantom{00} \\ \hline 80 \end{array}$$

12)
$$\begin{array}{r} 5 \\ - \phantom{0} \\ \hline 3 \end{array}$$

13)
$$\begin{array}{r} 69 \\ - \phantom{0} \\ \hline 8 \end{array}$$

14)
$$\begin{array}{r} 62 \\ - \phantom{00} \\ \hline 43 \end{array}$$

15)
$$\begin{array}{r} 83 \\ - \phantom{00} \\ \hline 34 \end{array}$$

16)
$$\begin{array}{r} 9 \\ - \phantom{0} \\ \hline 5 \end{array}$$

17)
$$\begin{array}{r} 40 \\ - \phantom{00} \\ \hline 12 \end{array}$$

18)
$$\begin{array}{r} 29 \\ - \phantom{00} \\ \hline 11 \end{array}$$

19)
$$\begin{array}{r} 29 \\ - \phantom{00} \\ \hline 16 \end{array}$$

20)
$$\begin{array}{r} 5 \\ - \phantom{0} \\ \hline 4 \end{array}$$

Let's sharpen our problem-solving skills by finding the missing numbers!

 **Name:** _____

 **Date:** _____

**Time Taken:** ___:___

 **Score:** ____ / **20**

**Day 53**

1)
```
    4
  - 
  ___
    3
```

2)
```
   50
  - 
  ___
   40
```

3)
```
   88
  - 
  ___
   46
```

4)
```
    6
  - 
  ___
```

5)
```
   36
  - 
  ___
    1
```

6)
```
   90
  - 
  ___
   82
```

7)
```
   50
  - 
  ___
   39
```

8)
```
    8
  - 
  ___
    7
```

9)
```
   74
  - 
  ___
   37
```

10)
```
   51
  - 
  ___
   17
```

11)
```
   21
  - 
  ___
   15
```

12)
```
    3
  - 
  ___
```

13)
```
   38
  - 
  ___
   29
```

14)
```
   20
  - 
  ___
    5
```

15)
```
   88
  - 
  ___
   36
```

16)
```
    7
  - 
  ___
    4
```

17)
```
   93
  - 
  ___
   81
```

18)
```
   85
  - 
  ___
   18
```

19)
```
   62
  - 
  ___
    0
```

20)
```
    6
  - 
  ___
    1
```

56

Missing numbers subtracting some clarity? Let's find the pieces!

 **Name:** _____

 **Time Taken:** ___:___

**Date:** _____

**Score:** / 20

**Day 54**

1)
```
  42
-
────
  38
```

2)
```
  82
-
────
  55
```

3)
```
  97
-
────
  45
```

4)
```
   7
-
────
   6
```

5)
```
  74
-
────
   9
```

6)
```
  28
-
────
  16
```

7)
```
  52
-
────
  44
```

8)
```
   4
-
────
   2
```

9)
```
  87
-
────
  83
```

10)
```
  76
-
────
  14
```

11)
```
  85
-
────
  79
```

12)
```
   5
-
────
   1
```

13)
```
  85
-
────
  77
```

14)
```
  95
-
────
  90
```

15)
```
  95
-
────
  24
```

16)
```
   6
-
────
```

17)
```
  89
-
────
  72
```

18)
```
  84
-
────
  54
```

19)
```
  93
-
────
  81
```

20)
```
   9
-
────
```

Ready to fill in the blanks and complete the subtraction puzzle? Let's do it!

 **Name:** _____

 **Date:** _____

**Day 55**

**Time Taken:** ___:___

 **Score:** ___ / 20

1)
$$\begin{array}{r} 90 \\ - \phantom{0} \\ \hline 5 \end{array}$$

2)
$$\begin{array}{r} 79 \\ - \phantom{0} \\ \hline 78 \end{array}$$

3)
$$\begin{array}{r} 51 \\ - \phantom{0} \\ \hline 13 \end{array}$$

4)
$$\begin{array}{r} 8 \\ - \phantom{0} \\ \hline 7 \end{array}$$

5)
$$\begin{array}{r} 65 \\ - \phantom{0} \\ \hline 45 \end{array}$$

6)
$$\begin{array}{r} 24 \\ - \phantom{0} \\ \hline 5 \end{array}$$

7)
$$\begin{array}{r} 59 \\ - \phantom{0} \\ \hline 0 \end{array}$$

8)
$$\begin{array}{r} 3 \\ - \phantom{0} \\ \hline \phantom{0} \end{array}$$

9)
$$\begin{array}{r} 94 \\ - \phantom{0} \\ \hline 55 \end{array}$$

10)
$$\begin{array}{r} 90 \\ - \phantom{0} \\ \hline 37 \end{array}$$

11)
$$\begin{array}{r} 53 \\ - \phantom{0} \\ \hline 26 \end{array}$$

12)
$$\begin{array}{r} 3 \\ - \phantom{0} \\ \hline 1 \end{array}$$

13)
$$\begin{array}{r} 73 \\ - \phantom{0} \\ \hline 2 \end{array}$$

14)
$$\begin{array}{r} 30 \\ - \phantom{0} \\ \hline 18 \end{array}$$

15)
$$\begin{array}{r} 51 \\ - \phantom{0} \\ \hline 40 \end{array}$$

16)
$$\begin{array}{r} 6 \\ - \phantom{0} \\ \hline 5 \end{array}$$

17)
$$\begin{array}{r} 25 \\ - \phantom{0} \\ \hline 21 \end{array}$$

18)
$$\begin{array}{r} 79 \\ - \phantom{0} \\ \hline 68 \end{array}$$

19)
$$\begin{array}{r} 31 \\ - \phantom{0} \\ \hline 10 \end{array}$$

20)
$$\begin{array}{r} 9 \\ - \phantom{0} \\ \hline 9 \end{array}$$

Let's put our detective hats on and hunt down the missing numbers!

**Day 56**

 **Name:** _____

**Time Taken:** ____:____

 **Date:** _____

 **Score:** _____ / **20**

1)
$$\begin{array}{r} 57 \\ - \phantom{00} \\ \hline 47 \end{array}$$

2)
$$\begin{array}{r} 3 \\ - \phantom{0} \\ \hline 1 \end{array}$$

3)
$$\begin{array}{r} 82 \\ - \phantom{00} \\ \hline 39 \end{array}$$

4)
$$\begin{array}{r} 9 \\ - \phantom{0} \\ \hline 4 \end{array}$$

5)
$$\begin{array}{r} 84 \\ - \phantom{00} \\ \hline 40 \end{array}$$

6)
$$\begin{array}{r} 40 \\ - \phantom{00} \\ \hline 16 \end{array}$$

7)
$$\begin{array}{r} 35 \\ - \phantom{00} \\ \hline 16 \end{array}$$

8)
$$\begin{array}{r} 4 \\ - \phantom{0} \\ \hline 3 \end{array}$$

9)
$$\begin{array}{r} 82 \\ - \phantom{00} \\ \hline 8 \end{array}$$

10)
$$\begin{array}{r} 45 \\ - \phantom{00} \\ \hline 8 \end{array}$$

11)
$$\begin{array}{r} 55 \\ - \phantom{00} \\ \hline 21 \end{array}$$

12)
$$\begin{array}{r} 5 \\ - \phantom{0} \\ \hline 2 \end{array}$$

13)
$$\begin{array}{r} 19 \\ - \phantom{00} \\ \hline 2 \end{array}$$

14)
$$\begin{array}{r} 98 \\ - \phantom{00} \\ \hline 85 \end{array}$$

15)
$$\begin{array}{r} 5 \\ - \phantom{0} \\ \hline 3 \end{array}$$

16)
$$\begin{array}{r} 5 \\ - \phantom{0} \\ \hline 4 \end{array}$$

17)
$$\begin{array}{r} 56 \\ - \phantom{00} \\ \hline 35 \end{array}$$

18)
$$\begin{array}{r} 63 \\ - \phantom{00} \\ \hline 40 \end{array}$$

19)
$$\begin{array}{r} 56 \\ - \phantom{00} \\ \hline 31 \end{array}$$

20)
$$\begin{array}{r} 4 \\ - \phantom{0} \\ \hline 2 \end{array}$$

Subtraction mysteries await! Let's uncover the missing pieces!

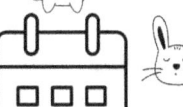

Name: _____

Date: _____

**Day 57**

Time Taken: ___:___

Score: ___ / 20

1)
$$\begin{array}{r} 34 \\ - \phantom{00} \\ \hline 12 \end{array}$$

2)
$$\begin{array}{r} 85 \\ - \phantom{00} \\ \hline 32 \end{array}$$

3)
$$\begin{array}{r} 65 \\ - \phantom{00} \\ \hline 34 \end{array}$$

4)
$$\begin{array}{r} 6 \\ - \phantom{00} \\ \hline 6 \end{array}$$

5)
$$\begin{array}{r} 91 \\ - \phantom{00} \\ \hline 80 \end{array}$$

6)
$$\begin{array}{r} 39 \\ - \phantom{00} \\ \hline 28 \end{array}$$

7)
$$\begin{array}{r} 53 \\ - \phantom{00} \\ \hline 3 \end{array}$$

8)
$$\begin{array}{r} 8 \\ - \phantom{00} \\ \hline \phantom{0} \end{array}$$

9)
$$\begin{array}{r} 87 \\ - \phantom{00} \\ \hline 83 \end{array}$$

10)
$$\begin{array}{r} 72 \\ - \phantom{00} \\ \hline 34 \end{array}$$

11)
$$\begin{array}{r} 70 \\ - \phantom{00} \\ \hline 46 \end{array}$$

12)
$$\begin{array}{r} 8 \\ - \phantom{00} \\ \hline 7 \end{array}$$

13)
$$\begin{array}{r} 56 \\ - \phantom{00} \\ \hline 1 \end{array}$$

14)
$$\begin{array}{r} 82 \\ - \phantom{00} \\ \hline 64 \end{array}$$

15)
$$\begin{array}{r} 72 \\ - \phantom{00} \\ \hline 41 \end{array}$$

16)
$$\begin{array}{r} 8 \\ - \phantom{00} \\ \hline 1 \end{array}$$

17)
$$\begin{array}{r} 46 \\ - \phantom{00} \\ \hline 6 \end{array}$$

18)
$$\begin{array}{r} 16 \\ - \phantom{00} \\ \hline 9 \end{array}$$

19)
$$\begin{array}{r} 78 \\ - \phantom{00} \\ \hline 41 \end{array}$$

20)
$$\begin{array}{r} 7 \\ - \phantom{00} \\ \hline 1 \end{array}$$

**Day 58**

Name: _____

Date: _____

Time Taken: ___:___

Score: / 20

1)
$$\begin{array}{r} 63 \\ - \phantom{00} \\ \hline 5 \end{array}$$

2)
$$\begin{array}{r} 43 \\ - \phantom{00} \\ \hline 34 \end{array}$$

3)
$$\begin{array}{r} 61 \\ - \phantom{00} \\ \hline 35 \end{array}$$

4)
$$\begin{array}{r} 7 \\ - \phantom{0} \\ \hline 1 \end{array}$$

5)
$$\begin{array}{r} 86 \\ - \phantom{00} \\ \hline 64 \end{array}$$

6)
$$\begin{array}{r} 55 \\ - \phantom{00} \\ \hline 40 \end{array}$$

7)
$$\begin{array}{r} 12 \\ - \phantom{00} \\ \hline 10 \end{array}$$

8)
$$\begin{array}{r} 6 \\ - \phantom{0} \\ \hline 4 \end{array}$$

9)
$$\begin{array}{r} 92 \\ - \phantom{00} \\ \hline 0 \end{array}$$

10)
$$\begin{array}{r} 88 \\ - \phantom{00} \\ \hline 43 \end{array}$$

11)
$$\begin{array}{r} 62 \\ - \phantom{00} \\ \hline 53 \end{array}$$

12)
$$\begin{array}{r} 9 \\ - \phantom{0} \\ \hline 9 \end{array}$$

13)
$$\begin{array}{r} 99 \\ - \phantom{00} \\ \hline 31 \end{array}$$

14)
$$\begin{array}{r} 92 \\ - \phantom{00} \\ \hline 40 \end{array}$$

15)
$$\begin{array}{r} 79 \\ - \phantom{00} \\ \hline 8 \end{array}$$

16)
$$\begin{array}{r} 5 \\ - \phantom{0} \\ \hline 4 \end{array}$$

17)
$$\begin{array}{r} 80 \\ - \phantom{00} \\ \hline 65 \end{array}$$

18)
$$\begin{array}{r} 69 \\ - \phantom{00} \\ \hline 63 \end{array}$$

19)
$$\begin{array}{r} 79 \\ - \phantom{00} \\ \hline 71 \end{array}$$

20)
$$\begin{array}{r} 9 \\ - \phantom{0} \\ \hline 7 \end{array}$$

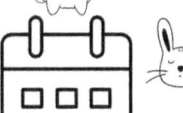

**Name:** _____

**Date:** _____

**Time Taken:** ___:___

**Score:** ___ / 20

**Day 59**

1)
```
   52
 -
 ────
   49
```

2)
```
   99
 -
 ────
   75
```

3)
```
   83
 -
 ────
   52
```

4)
```
    7
 -
 ────
    5
```

5)
```
   73
 -
 ────
    8
```

6)
```
   84
 -
 ────
   81
```

7)
```
   35
 -
 ────
   30
```

8)
```
    8
 -
 ────
    3
```

9)
```
   92
 -
 ────
   77
```

10)
```
   37
 -
 ────
    9
```

11)
```
   27
 -
 ────
    8
```

12)
```
    5
 -
 ────
    4
```

13)
```
   22
 -
 ────
   19
```

14)
```
   32
 -
 ────
   23
```

15)
```
   93
 -
 ────
   51
```

16)
```
    9
 -
 ────
    5
```

17)
```
   52
 -
 ────
   36
```

18)
```
   92
 -
 ────
   29
```

19)
```
   31
 -
 ────
    8
```

20)
```
    9
 -
 ────
    6
```

Time to find the missing links in our subtraction equations. Let's connect the dots!

 **Name:** _____        **Date:** _____

**Day 60** **Time Taken:** ___:___        **Score:** ___ / **20**

1)
$$98$$
$$-$$
$$9$$

2)
$$14$$
$$-$$
$$7$$

3)
$$77$$
$$-$$
$$28$$

4)
$$6$$
$$-$$
$$1$$

5)
$$79$$
$$-$$
$$17$$

6)
$$56$$
$$-$$
$$31$$

7)
$$95$$
$$-$$
$$15$$

8)
$$1$$
$$-$$
$$1$$

9)
$$63$$
$$-$$
$$28$$

10)
$$51$$
$$-$$
$$42$$

11)
$$32$$
$$-$$
$$28$$

12)
$$6$$
$$-$$
$$4$$

13)
$$27$$
$$-$$
$$25$$

14)
$$71$$
$$-$$
$$34$$

15)
$$76$$
$$-$$
$$58$$

16)
$$8$$
$$-$$
$$5$$

17)
$$79$$
$$-$$
$$63$$

18)
$$95$$
$$-$$
$$58$$

19)
$$78$$
$$-$$
$$47$$

20)
$$1$$
$$-$$

 Let's multiply and make some big numbers! Ready?

1)
```
    81
×   89
```

2)
```
    91
×   57
```

3)
```
    96
×   95
```

4)
```
    94
×   31
```

5)
```
    59
×   78
```

6)
```
    68
×   31
```

7)
```
    28
×   97
```

8)
```
    12
×    4
```

9)
```
    64
×   33
```

10)
```
    20
×   15
```

11)
```
    94
×   50
```

12)
```
    93
×   90
```

13)
```
    21
×   28
```

14)
```
    99
×   48
```

15)
```
    11
×   27
```

16)
```
    14
×   96
```

17)
```
    57
×   10
```

18)
```
    31
×   92
```

19)
```
    13
×   28
```

20)
```
    89
×   19
```

**Name:** _____

**Date:** _____

**Time Taken:** ___:___

**Score:** / 20

**Day 62**

1)
$$\begin{array}{r} 71 \\ \times\ 94 \\ \hline \end{array}$$

2)
$$\begin{array}{r} 43 \\ \times\ 24 \\ \hline \end{array}$$

3)
$$\begin{array}{r} 85 \\ \times\ 80 \\ \hline \end{array}$$

4)
$$\begin{array}{r} 93 \\ \times\ 38 \\ \hline \end{array}$$

5)
$$\begin{array}{r} 69 \\ \times\ 70 \\ \hline \end{array}$$

6)
$$\begin{array}{r} 45 \\ \times\ 98 \\ \hline \end{array}$$

7)
$$\begin{array}{r} 50 \\ \times\ 77 \\ \hline \end{array}$$

8)
$$\begin{array}{r} 76 \\ \times\ 22 \\ \hline \end{array}$$

9)
$$\begin{array}{r} 78 \\ \times\ 85 \\ \hline \end{array}$$

10)
$$\begin{array}{r} 28 \\ \times\ 95 \\ \hline \end{array}$$

11)
$$\begin{array}{r} 19 \\ \times\ 30 \\ \hline \end{array}$$

12)
$$\begin{array}{r} 36 \\ \times\ 20 \\ \hline \end{array}$$

13)
$$\begin{array}{r} 6 \\ \times\ 84 \\ \hline \end{array}$$

14)
$$\begin{array}{r} 95 \\ \times\ 64 \\ \hline \end{array}$$

15)
$$\begin{array}{r} 11 \\ \times\ 58 \\ \hline \end{array}$$

16)
$$\begin{array}{r} 34 \\ \times\ 75 \\ \hline \end{array}$$

17)
$$\begin{array}{r} 39 \\ \times\ 78 \\ \hline \end{array}$$

18)
$$\begin{array}{r} 25 \\ \times\ 7 \\ \hline \end{array}$$

19)
$$\begin{array}{r} 53 \\ \times\ 61 \\ \hline \end{array}$$

20)
$$\begin{array}{r} 80 \\ \times\ 28 \\ \hline \end{array}$$

**Day 63**

 **Name:** _____

 **Time Taken:** ___:___

**Date:** _____

**Score:** / **20**

1)
```
    99
×   21
```

2)
```
    36
×   37
```

3)
```
    13
×   52
```

4)
```
    72
×   88
```

5)
```
    43
×    8
```

6)
```
     5
×   46
```

7)
```
    52
×   33
```

8)
```
    26
×   22
```

9)
```
    77
×   63
```

10)
```
    11
×   65
```

11)
```
     6
×   37
```

12)
```
    30
×   68
```

13)
```
    45
×    3
```

14)
```
    36
×   88
```

15)
```
    55
×    7
```

16)
```
    96
×   76
```

17)
```
    95
×   15
```

18)
```
    64
×   55
```

19)
```
    99
×   83
```

20)
```
    61
×   29
```

**Day 64**

Name: _____

Time Taken: ___:___

Date: _____

Score: / 20

1)  16
  × 37

2)  88
  × 22

3)  55
  × 41

4)  98
  ×  2

5)  51
  × 31

6)  50
  × 98

7)  61
  × 51

8)  79
  × 53

9)  50
  × 67

10)  48
  × 35

11)  59
  × 54

12)  85
  × 42

13)  49
  × 54

14)  63
  × 59

15)  34
  × 30

16)  61
  × 78

17)  64
  ×  1

18)  20
  × 21

19)  67
  × 66

20)  23
  × 93

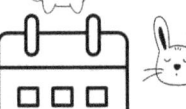

**Name:** _____

**Date:** _____

**Day 65**

**Time Taken:** ___:___

**Score:** / 20

1)
$$\begin{array}{r} 12 \\ \times\ 12 \\ \hline \end{array}$$

2)
$$\begin{array}{r} 22 \\ \times\ 94 \\ \hline \end{array}$$

3)
$$\begin{array}{r} 47 \\ \times\ 37 \\ \hline \end{array}$$

4)
$$\begin{array}{r} 54 \\ \times\ 69 \\ \hline \end{array}$$

5)
$$\begin{array}{r} 61 \\ \times\ 81 \\ \hline \end{array}$$

6)
$$\begin{array}{r} 43 \\ \times\ 99 \\ \hline \end{array}$$

7)
$$\begin{array}{r} 71 \\ \times\ 70 \\ \hline \end{array}$$

8)
$$\begin{array}{r} 47 \\ \times\ 54 \\ \hline \end{array}$$

9)
$$\begin{array}{r} 9 \\ \times\ 90 \\ \hline \end{array}$$

10)
$$\begin{array}{r} 17 \\ \times\ 84 \\ \hline \end{array}$$

11)
$$\begin{array}{r} 60 \\ \times\ 73 \\ \hline \end{array}$$

12)
$$\begin{array}{r} 46 \\ \times\ 67 \\ \hline \end{array}$$

13)
$$\begin{array}{r} 97 \\ \times\ 83 \\ \hline \end{array}$$

14)
$$\begin{array}{r} 67 \\ \times\ 14 \\ \hline \end{array}$$

15)
$$\begin{array}{r} 39 \\ \times\ 35 \\ \hline \end{array}$$

16)
$$\begin{array}{r} 33 \\ \times\ 13 \\ \hline \end{array}$$

17)
$$\begin{array}{r} 21 \\ \times\ 21 \\ \hline \end{array}$$

18)
$$\begin{array}{r} 98 \\ \times\ 40 \\ \hline \end{array}$$

19)
$$\begin{array}{r} 71 \\ \times\ 99 \\ \hline \end{array}$$

20)
$$\begin{array}{r} 64 \\ \times\ 91 \\ \hline \end{array}$$

Let's activate our multiplication powers. Multiplication mode: on!

**Day 66**

🐰 **Name:** _____

⏱ **Time Taken:** ___:___

📅 **Date:** _____

📋 **Score:** / 20

1)
$$\begin{array}{r} 14 \\ \times\ 61 \\ \hline \end{array}$$

2)
$$\begin{array}{r} 73 \\ \times\ 5 \\ \hline \end{array}$$

3)
$$\begin{array}{r} 63 \\ \times\ 78 \\ \hline \end{array}$$

4)
$$\begin{array}{r} 68 \\ \times\ 65 \\ \hline \end{array}$$

5)
$$\begin{array}{r} 80 \\ \times\ 23 \\ \hline \end{array}$$

6)
$$\begin{array}{r} 58 \\ \times\ 46 \\ \hline \end{array}$$

7)
$$\begin{array}{r} 1 \\ \times\ 6 \\ \hline \end{array}$$

8)
$$\begin{array}{r} 3 \\ \times\ 64 \\ \hline \end{array}$$

9)
$$\begin{array}{r} 20 \\ \times\ 70 \\ \hline \end{array}$$

10)
$$\begin{array}{r} 20 \\ \times\ 50 \\ \hline \end{array}$$

11)
$$\begin{array}{r} 43 \\ \times\ 83 \\ \hline \end{array}$$

12)
$$\begin{array}{r} 82 \\ \times\ 45 \\ \hline \end{array}$$

13)
$$\begin{array}{r} 46 \\ \times\ 24 \\ \hline \end{array}$$

14)
$$\begin{array}{r} 30 \\ \times\ 69 \\ \hline \end{array}$$

15)
$$\begin{array}{r} 47 \\ \times\ 81 \\ \hline \end{array}$$

16)
$$\begin{array}{r} 83 \\ \times\ 17 \\ \hline \end{array}$$

17)
$$\begin{array}{r} 44 \\ \times\ 12 \\ \hline \end{array}$$

18)
$$\begin{array}{r} 58 \\ \times\ 86 \\ \hline \end{array}$$

19)
$$\begin{array}{r} 57 \\ \times\ 99 \\ \hline \end{array}$$

20)
$$\begin{array}{r} 33 \\ \times\ 84 \\ \hline \end{array}$$

 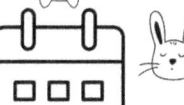

Multiplying numbers like a pro! Let's tackle these equations!

 **Name:** _____

 **Date:** _____

**Day 67**

**Time Taken:** ___:___

 **Score:** / **20**

1)
$$93 \times 8$$

2)
$$74 \times 19$$

3)
$$67 \times 52$$

4)
$$5 \times 39$$

5)
$$81 \times 98$$

6)
$$13 \times 22$$

7)
$$65 \times 30$$

8)
$$40 \times 59$$

9)
$$50 \times 88$$

10)
$$95 \times 48$$

11)
$$95 \times 78$$

12)
$$99 \times 22$$

13)
$$83 \times 9$$

14)
$$90 \times 53$$

15)
$$37 \times 43$$

16)
$$76 \times 9$$

17)
$$62 \times 84$$

18)
$$91 \times 99$$

19)
$$34 \times 56$$

20)
$$95 \times 52$$

Ready to multiply and amplify our math skills? Let's go!

1)  85
   × 58

2)  74
   × 46

3)   4
   × 1

4)  92
   × 1

5)   2
   × 36

6)  14
   × 96

7)  18
   × 50

8)  67
   × 34

9)   2
   × 70

10)   6
    × 74

11)  11
    × 75

12)  61
    × 21

13)  46
    × 47

14)  18
    × 75

15)  76
    × 64

16)  23
    × 91

17)  78
    × 36

18)  79
    × 10

19)  11
    × 65

20)  40
    × 77

Let's multiply the excitement with some multiplication challenges!

1)
$$\begin{array}{r} 25 \\ \times\ 49 \\ \hline \end{array}$$

2)
$$\begin{array}{r} 44 \\ \times\ 50 \\ \hline \end{array}$$

3)
$$\begin{array}{r} 43 \\ \times\ 98 \\ \hline \end{array}$$

4)
$$\begin{array}{r} 63 \\ \times\ 80 \\ \hline \end{array}$$

5)
$$\begin{array}{r} 59 \\ \times\ 13 \\ \hline \end{array}$$

6)
$$\begin{array}{r} 87 \\ \times\ 15 \\ \hline \end{array}$$

7)
$$\begin{array}{r} 2 \\ \times\ 13 \\ \hline \end{array}$$

8)
$$\begin{array}{r} 57 \\ \times\ 92 \\ \hline \end{array}$$

9)
$$\begin{array}{r} 48 \\ \times\ 27 \\ \hline \end{array}$$

10)
$$\begin{array}{r} 84 \\ \times\ 48 \\ \hline \end{array}$$

11)
$$\begin{array}{r} 88 \\ \times\ 90 \\ \hline \end{array}$$

12)
$$\begin{array}{r} 2 \\ \times\ 87 \\ \hline \end{array}$$

13)
$$\begin{array}{r} 52 \\ \times\ 99 \\ \hline \end{array}$$

14)
$$\begin{array}{r} 55 \\ \times\ 14 \\ \hline \end{array}$$

15)
$$\begin{array}{r} 49 \\ \times\ 11 \\ \hline \end{array}$$

16)
$$\begin{array}{r} 48 \\ \times\ 74 \\ \hline \end{array}$$

17)
$$\begin{array}{r} 30 \\ \times\ 30 \\ \hline \end{array}$$

18)
$$\begin{array}{r} 28 \\ \times\ 96 \\ \hline \end{array}$$

19)
$$\begin{array}{r} 57 \\ \times\ 61 \\ \hline \end{array}$$

20)
$$\begin{array}{r} 73 \\ \times\ 39 \\ \hline \end{array}$$

 **Name:** _____

 **Date:** _____

**Day 70**

**Time Taken:** \_\_\_:\_\_\_

 **Score:** / **20**

1)
$$\begin{array}{r} 3 \\ \times\ 57 \\ \hline \end{array}$$

2)
$$\begin{array}{r} 95 \\ \times\ 93 \\ \hline \end{array}$$

3)
$$\begin{array}{r} 77 \\ \times\ 67 \\ \hline \end{array}$$

4)
$$\begin{array}{r} 48 \\ \times\ 70 \\ \hline \end{array}$$

5)
$$\begin{array}{r} 65 \\ \times\ 92 \\ \hline \end{array}$$

6)
$$\begin{array}{r} 69 \\ \times\ 81 \\ \hline \end{array}$$

7)
$$\begin{array}{r} 68 \\ \times\ 54 \\ \hline \end{array}$$

8)
$$\begin{array}{r} 77 \\ \times\ 75 \\ \hline \end{array}$$

9)
$$\begin{array}{r} 94 \\ \times\ 14 \\ \hline \end{array}$$

10)
$$\begin{array}{r} 22 \\ \times\ 59 \\ \hline \end{array}$$

11)
$$\begin{array}{r} 31 \\ \times\ 22 \\ \hline \end{array}$$

12)
$$\begin{array}{r} 36 \\ \times\ 64 \\ \hline \end{array}$$

13)
$$\begin{array}{r} 95 \\ \times\ 57 \\ \hline \end{array}$$

14)
$$\begin{array}{r} 36 \\ \times\ 17 \\ \hline \end{array}$$

15)
$$\begin{array}{r} 26 \\ \times\ 8 \\ \hline \end{array}$$

16)
$$\begin{array}{r} 37 \\ \times\ 84 \\ \hline \end{array}$$

17)
$$\begin{array}{r} 31 \\ \times\ 68 \\ \hline \end{array}$$

18)
$$\begin{array}{r} 83 \\ \times\ 87 \\ \hline \end{array}$$

19)
$$\begin{array}{r} 56 \\ \times\ 98 \\ \hline \end{array}$$

20)
$$\begin{array}{r} 89 \\ \times\ 55 \\ \hline \end{array}$$

**Day 71**

Name: _____

Time Taken: ___:___

Date: _____

Score: / 20

1) $63 \times 9 =$

2) $4 \times 7 =$

3) $36 \times 4 =$

4) $5 \times 86 =$

5) $7 \times 85 =$

6) $8 \times 42 =$

7) $8 \times 85 =$

8) $3 \times 55 =$

9) $5 \times 82 =$

10) $97 \times 5 =$

11) $3 \times 84 =$

12) $76 \times 8 =$

13) $8 \times 3 =$

14) $4 \times 25 =$

15) $68 \times 3 =$

16) $59 \times 6 =$

17) $16 \times 3 =$

18) $25 \times 8 =$

19) $15 \times 7 =$

20) $3 \times 9 =$

 **Name:** _____

 **Date:** _____

**Day 72**  **Time Taken:** ___:___

**Score:** ___ / **20**

1) $2 \times 31 =$

2) $9 \times 92 =$

3) $8 \times 3 =$

4) $69 \times 8 =$

5) $7 \times 79 =$

6) $6 \times 32 =$

7) $59 \times 2 =$

8) $3 \times 29 =$

9) $68 \times 3 =$

10) $9 \times 43 =$

11) $27 \times 7 =$

12) $26 \times 6 =$

13) $55 \times 8 =$

14) $64 \times 4 =$

15) $6 \times 36 =$

16) $5 \times 10 =$

17) $2 \times 72 =$

18) $52 \times 5 =$

19) $24 \times 9 =$

20) $34 \times 5 =$

Let's sharpen our mental math skills with some quick multiplication!

  **Name:** _____  **Date:** _____

**Day 73**  **Time Taken:** ___:___  **Score:** / **20**

1) 6 × 6 =

2) 8 × 61 =

3) 3 × 70 =

4) 5 × 52 =

5) 2 × 37 =

6) 5 × 43 =

7) 98 × 5 =

8) 5 × 15 =

9) 11 × 2 =

10) 47 × 7 =

11) 9 × 65 =

12) 97 × 5 =

13) 65 × 2 =

14) 7 × 74 =

15) 7 × 21 =

16) 8 × 56 =

17) 14 × 8 =

18) 96 × 7 =

19) 6 × 28 =

20) 8 × 70 =

 Mental math magic time! Let's multiply with speed and accuracy!

1) 6 × 48 =

2) 85 × 9 =

3) 83 × 4 =

4) 38 × 3 =

5) 3 × 67 =

6) 81 × 7 =

7) 2 × 6 =

8) 97 × 6 =

9) 79 × 7 =

10) 75 × 2 =

11) 9 × 91 =

12) 19 × 5 =

13) 9 × 64 =

14) 21 × 3 =

15) 77 × 9 =

16) 8 × 52 =

17) 78 × 3 =

18) 6 × 4 =

19) 3 × 79 =

20) 51 × 6 =

**Day 75**

Name: _____

Date: _____

Time Taken: ___:___

Score: / 20

1) $91 \times 9 =$

2) $2 \times 14 =$

3) $6 \times 10 =$

4) $4 \times 4 =$

5) $2 \times 44 =$

6) $33 \times 8 =$

7) $2 \times 59 =$

8) $3 \times 23 =$

9) $4 \times 78 =$

10) $38 \times 9 =$

11) $6 \times 62 =$

12) $47 \times 5 =$

13) $2 \times 23 =$

14) $6 \times 87 =$

15) $2 \times 56 =$

16) $5 \times 63 =$

17) $8 \times 65 =$

18) $3 \times 88 =$

19) $30 \times 7 =$

20) $3 \times 38 =$

**Name:** _____

**Date:** _____

**Day 76**

**Time Taken:** ___:___

**Score:** ___ / 20

1) 9 × 36 =

2) 9 × 54 =

3) 84 × 6 =

4) 9 × 73 =

5) 27 × 6 =

6) 81 × 9 =

7) 23 × 4 =

8) 4 × 7 =

9) 4 × 34 =

10) 5 × 5 =

11) 79 × 6 =

12) 2 × 69 =

13) 98 × 2 =

14) 2 × 47 =

15) 6 × 76 =

16) 8 × 56 =

17) 49 × 9 =

18) 2 × 27 =

19) 3 × 30 =

20) 4 × 86 =

**Day 77**

 **Name:** _____

 **Time Taken:** ___:___

 **Date:** _____

**Score:** / 20

1) 4 × 64 =

2) 52 × 3 =

3) 7 × 27 =

4) 7 × 37 =

5) 78 × 6 =

6) 5 × 78 =

7) 85 × 9 =

8) 4 × 65 =

9) 5 × 9 =

10) 4 × 96 =

11) 95 × 9 =

12) 8 × 39 =

13) 95 × 3 =

14) 48 × 4 =

15) 8 × 13 =

16) 3 × 48 =

17) 7 × 42 =

18) 5 × 62 =

19) 97 × 8 =

20) 8 × 71 =

Let's activate our mental math powers. Multiplication mode: on!

 **Name:** _____

 **Date:** _____

**Day 78**

 **Time Taken:** ___:___

 **Score:** / **20**

1) 4 × 76 =

2) 80 × 3 =

3) 51 × 7 =

4) 2 × 67 =

5) 46 × 4 =

6) 8 × 73 =

7) 3 × 64 =

8) 35 × 6 =

9) 8 × 21 =

10) 27 × 3 =

11) 5 × 64 =

12) 9 × 90 =

13) 7 × 39 =

14) 49 × 2 =

15) 32 × 5 =

16) 45 × 4 =

17) 34 × 3 =

18) 4 × 59 =

19) 27 × 8 =

20) 4 × 95 =

Mental math wizards, unite! Let's tackle these multiplication problems!

 **Name:** _____

 **Date:** _____

**Time Taken:** ___:___

**Score:** / 20

1) 75 × 9 =

2) 7 × 38 =

3) 2 × 13 =

4) 28 × 4 =

5) 48 × 3 =

6) 9 × 35 =

7) 6 × 33 =

8) 4 × 8 =

9) 8 × 22 =

10) 81 × 7 =

11) 54 × 5 =

12) 5 × 7 =

13) 2 × 62 =

14) 64 × 9 =

15) 89 × 4 =

16) 89 × 6 =

17) 17 × 3 =

18) 29 × 4 =

19) 93 × 8 =

20) 4 × 7 =

**Name:** _____

**Date:** _____

**Time Taken:** ___:___

**Score:** / 20

**Day 80**

1) 2 × 56 =

2) 8 × 16 =

3) 96 × 4 =

4) 3 × 22 =

5) 7 × 87 =

6) 2 × 97 =

7) 97 × 9 =

8) 77 × 2 =

9) 8 × 87 =

10) 9 × 59 =

11) 10 × 5 =

12) 6 × 15 =

13) 61 × 2 =

14) 4 × 88 =

15) 53 × 7 =

16) 9 × 52 =

17) 74 × 2 =

18) 5 × 13 =

19) 54 × 9 =

20) 6 × 62 =

Can you fill in the missing pieces? Let's solve some multiplication puzzles!

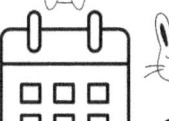

 **Name:** _____

 **Date:** _____

**Day 81**   **Time Taken:** ___:___

**Score:**  / 20

1)
$$
\begin{array}{r}
36 \\
\times \phantom{00} \\
\hline
216
\end{array}
$$

2)
$$
\begin{array}{r}
15 \\
\times \phantom{00} \\
\hline
90
\end{array}
$$

3)
$$
\begin{array}{r}
74 \\
\times \phantom{00} \\
\hline
592
\end{array}
$$

4)
$$
\begin{array}{r}
5 \\
\times \phantom{00} \\
\hline
10
\end{array}
$$

5)
$$
\begin{array}{r}
34 \\
\times \phantom{00} \\
\hline
204
\end{array}
$$

6)
$$
\begin{array}{r}
95 \\
\times \phantom{00} \\
\hline
190
\end{array}
$$

7)
$$
\begin{array}{r}
37 \\
\times \phantom{00} \\
\hline
259
\end{array}
$$

8)
$$
\begin{array}{r}
3 \\
\times \phantom{00} \\
\hline
26
\end{array}
$$

9)
$$
\begin{array}{r}
2 \\
\times \phantom{00} \\
\hline
16
\end{array}
$$

10)
$$
\begin{array}{r}
18 \\
\times \phantom{00} \\
\hline
126
\end{array}
$$

11)
$$
\begin{array}{r}
56 \\
\times \phantom{00} \\
\hline
168
\end{array}
$$

12)
$$
\begin{array}{r}
6 \\
\times \phantom{00} \\
\hline
36
\end{array}
$$

13)
$$
\begin{array}{r}
31 \\
\times \phantom{00} \\
\hline
62
\end{array}
$$

14)
$$
\begin{array}{r}
42 \\
\times \phantom{00} \\
\hline
210
\end{array}
$$

15)
$$
\begin{array}{r}
33 \\
\times \phantom{00} \\
\hline
66
\end{array}
$$

16)
$$
\begin{array}{r}
6 \\
\times \phantom{00} \\
\hline
27
\end{array}
$$

17)
$$
\begin{array}{r}
25 \\
\times \phantom{00} \\
\hline
150
\end{array}
$$

18)
$$
\begin{array}{r}
94 \\
\times \phantom{00} \\
\hline
188
\end{array}
$$

19)
$$
\begin{array}{r}
40 \\
\times \phantom{00} \\
\hline
120
\end{array}
$$

20)
$$
\begin{array}{r}
2 \\
\times \phantom{00} \\
\hline
8
\end{array}
$$

 **Name:** _____   **Date:** _____

**Day 82**  **Time Taken:** ___:___  **Score:**  / 20

1)
$$\begin{array}{r} 20 \\ \times \phantom{00} \\ \hline 100 \end{array}$$

2)
$$\begin{array}{r} 92 \\ \times \phantom{00} \\ \hline 736 \end{array}$$

3)
$$\begin{array}{r} 41 \\ \times \phantom{00} \\ \hline 328 \end{array}$$

4)
$$\begin{array}{r} 2 \\ \times \phantom{00} \\ \hline 20 \end{array}$$

5)
$$\begin{array}{r} 16 \\ \times \phantom{00} \\ \hline 128 \end{array}$$

6)
$$\begin{array}{r} 60 \\ \times \phantom{00} \\ \hline 240 \end{array}$$

7)
$$\begin{array}{r} 35 \\ \times \phantom{00} \\ \hline 105 \end{array}$$

8)
$$\begin{array}{r} 6 \\ \times \phantom{00} \\ \hline 19 \end{array}$$

9)
$$\begin{array}{r} 38 \\ \times \phantom{00} \\ \hline 266 \end{array}$$

10)
$$\begin{array}{r} 22 \\ \times \phantom{00} \\ \hline 176 \end{array}$$

11)
$$\begin{array}{r} 98 \\ \times \phantom{00} \\ \hline 294 \end{array}$$

12)
$$\begin{array}{r} 7 \\ \times \phantom{00} \\ \hline 14 \end{array}$$

13)
$$\begin{array}{r} 10 \\ \times \phantom{00} \\ \hline 60 \end{array}$$

14)
$$\begin{array}{r} 85 \\ \times \phantom{00} \\ \hline 595 \end{array}$$

15)
$$\begin{array}{r} 24 \\ \times \phantom{00} \\ \hline 120 \end{array}$$

16)
$$\begin{array}{r} 7 \\ \times \phantom{00} \\ \hline 36 \end{array}$$

17)
$$\begin{array}{r} 53 \\ \times \phantom{00} \\ \hline 159 \end{array}$$

18)
$$\begin{array}{r} 71 \\ \times \phantom{00} \\ \hline 568 \end{array}$$

19)
$$\begin{array}{r} 90 \\ \times \phantom{00} \\ \hline 720 \end{array}$$

20)
$$\begin{array}{r} 1 \\ \times \phantom{00} \\ \hline 15 \end{array}$$

Missing numbers multiplying our equations? Let's complete the picture!

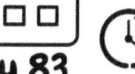 **Day 83**

 **Name:** _____

**Time Taken:** ___:___

**Date:** _____

**Score:** / 20

1)  $\begin{array}{r} 44 \\ \times \underline{\phantom{00}} \\ 220 \end{array}$

2)  $\begin{array}{r} 45 \\ \times \underline{\phantom{00}} \\ 360 \end{array}$

3)  $\begin{array}{r} 35 \\ \times \underline{\phantom{00}} \\ 245 \end{array}$

4)  $\begin{array}{r} 8 \\ \times \underline{\phantom{00}} \\ 52 \end{array}$

5)  $\begin{array}{r} 61 \\ \times \underline{\phantom{00}} \\ 122 \end{array}$

6)  $\begin{array}{r} 95 \\ \times \underline{\phantom{00}} \\ 665 \end{array}$

7)  $\begin{array}{r} 28 \\ \times \underline{\phantom{00}} \\ 224 \end{array}$

8)  $\begin{array}{r} 3 \\ \times \underline{\phantom{00}} \\ 9 \end{array}$

9)  $\begin{array}{r} 76 \\ \times \underline{\phantom{00}} \\ 532 \end{array}$

10) $\begin{array}{r} 94 \\ \times \underline{\phantom{00}} \\ 470 \end{array}$

11) $\begin{array}{r} 3 \\ \times \underline{\phantom{00}} \\ 12 \end{array}$

12) $\begin{array}{r} 5 \\ \times \underline{\phantom{00}} \\ 51 \end{array}$

13) $\begin{array}{r} 12 \\ \times \underline{\phantom{00}} \\ 108 \end{array}$

14) $\begin{array}{r} 37 \\ \times \underline{\phantom{00}} \\ 222 \end{array}$

15) $\begin{array}{r} 63 \\ \times \underline{\phantom{00}} \\ 441 \end{array}$

16) $\begin{array}{r} 6 \\ \times \underline{\phantom{00}} \\ 19 \end{array}$

17) $\begin{array}{r} 45 \\ \times \underline{\phantom{00}} \\ 180 \end{array}$

18) $\begin{array}{r} 84 \\ \times \underline{\phantom{00}} \\ 420 \end{array}$

19) $\begin{array}{r} 47 \\ \times \underline{\phantom{00}} \\ 423 \end{array}$

20) $\begin{array}{r} 3 \\ \times \underline{\phantom{00}} \\ 12 \end{array}$

 Let's activate our problem-solving skills by filling in the missing numbers!

1)
$$
\begin{array}{r}
15 \\
\times \underline{\phantom{00}} \\
105
\end{array}
$$

2)
$$
\begin{array}{r}
21 \\
\times \underline{\phantom{00}} \\
105
\end{array}
$$

3)
$$
\begin{array}{r}
37 \\
\times \underline{\phantom{00}} \\
296
\end{array}
$$

4)
$$
\begin{array}{r}
1 \\
\times \underline{\phantom{00}} \\
9
\end{array}
$$

5)
$$
\begin{array}{r}
44 \\
\times \underline{\phantom{00}} \\
132
\end{array}
$$

6)
$$
\begin{array}{r}
46 \\
\times \underline{\phantom{00}} \\
138
\end{array}
$$

7)
$$
\begin{array}{r}
24 \\
\times \underline{\phantom{00}} \\
192
\end{array}
$$

8)
$$
\begin{array}{r}
9 \\
\times \underline{\phantom{00}} \\
55
\end{array}
$$

9)
$$
\begin{array}{r}
18 \\
\times \underline{\phantom{00}} \\
36
\end{array}
$$

10)
$$
\begin{array}{r}
96 \\
\times \underline{\phantom{00}} \\
768
\end{array}
$$

11)
$$
\begin{array}{r}
40 \\
\times \underline{\phantom{00}} \\
320
\end{array}
$$

12)
$$
\begin{array}{r}
8 \\
\times \underline{\phantom{00}} \\
26
\end{array}
$$

13)
$$
\begin{array}{r}
11 \\
\times \underline{\phantom{00}} \\
88
\end{array}
$$

14)
$$
\begin{array}{r}
23 \\
\times \underline{\phantom{00}} \\
115
\end{array}
$$

15)
$$
\begin{array}{r}
88 \\
\times \underline{\phantom{00}} \\
792
\end{array}
$$

16)
$$
\begin{array}{r}
4 \\
\times \underline{\phantom{00}} \\
9
\end{array}
$$

17)
$$
\begin{array}{r}
12 \\
\times \underline{\phantom{00}} \\
48
\end{array}
$$

18)
$$
\begin{array}{r}
26 \\
\times \underline{\phantom{00}} \\
208
\end{array}
$$

19)
$$
\begin{array}{r}
43 \\
\times \underline{\phantom{00}} \\
86
\end{array}
$$

20)
$$
\begin{array}{r}
8 \\
\times \underline{\phantom{00}} \\
24
\end{array}
$$

Ready to put our detective hats on and uncover the missing numbers?

1)
$$\begin{array}{r} 37 \\ \times \phantom{00} \\ \hline 148 \end{array}$$

2)
$$\begin{array}{r} 60 \\ \times \phantom{00} \\ \hline 120 \end{array}$$

3)
$$\begin{array}{r} 13 \\ \times \phantom{00} \\ \hline 65 \end{array}$$

4)
$$\begin{array}{r} 6 \\ \times \phantom{00} \\ \hline 51 \end{array}$$

5)
$$\begin{array}{r} 29 \\ \times \phantom{00} \\ \hline 87 \end{array}$$

6)
$$\begin{array}{r} 53 \\ \times \phantom{00} \\ \hline 424 \end{array}$$

7)
$$\begin{array}{r} 34 \\ \times \phantom{00} \\ \hline 102 \end{array}$$

8)
$$\begin{array}{r} 8 \\ \times \phantom{00} \\ \hline 65 \end{array}$$

9)
$$\begin{array}{r} 19 \\ \times \phantom{00} \\ \hline 152 \end{array}$$

10)
$$\begin{array}{r} 8 \\ \times \phantom{00} \\ \hline 48 \end{array}$$

11)
$$\begin{array}{r} 89 \\ \times \phantom{00} \\ \hline 445 \end{array}$$

12)
$$\begin{array}{r} 3 \\ \times \phantom{00} \\ \hline 24 \end{array}$$

13)
$$\begin{array}{r} 7 \\ \times \phantom{00} \\ \hline 42 \end{array}$$

14)
$$\begin{array}{r} 2 \\ \times \phantom{00} \\ \hline 16 \end{array}$$

15)
$$\begin{array}{r} 53 \\ \times \phantom{00} \\ \hline 159 \end{array}$$

16)
$$\begin{array}{r} 2 \\ \times \phantom{00} \\ \hline 23 \end{array}$$

17)
$$\begin{array}{r} 47 \\ \times \phantom{00} \\ \hline 329 \end{array}$$

18)
$$\begin{array}{r} 25 \\ \times \phantom{00} \\ \hline 125 \end{array}$$

19)
$$\begin{array}{r} 48 \\ \times \phantom{00} \\ \hline 336 \end{array}$$

20)
$$\begin{array}{r} 1 \\ \times \phantom{00} \\ \hline 2 \end{array}$$

Let's sharpen our multiplication skills by finding the missing numbers!

 **Name:** _____    **Date:** _____

**Day 86**  **Time Taken:** ___:___   **Score:**   /  **20**

1)
$$
\begin{array}{r}
44 \\
\times \underline{\phantom{00}} \\
352
\end{array}
$$

2)
$$
\begin{array}{r}
84 \\
\times \underline{\phantom{00}} \\
672
\end{array}
$$

3)
$$
\begin{array}{r}
11 \\
\times \underline{\phantom{00}} \\
22
\end{array}
$$

4)
$$
\begin{array}{r}
5 \\
\times \underline{\phantom{00}} \\
43
\end{array}
$$

5)
$$
\begin{array}{r}
99 \\
\times \underline{\phantom{00}} \\
594
\end{array}
$$

6)
$$
\begin{array}{r}
84 \\
\times \underline{\phantom{00}} \\
420
\end{array}
$$

7)
$$
\begin{array}{r}
66 \\
\times \underline{\phantom{00}} \\
198
\end{array}
$$

8)
$$
\begin{array}{r}
1 \\
\times \underline{\phantom{00}} \\
3
\end{array}
$$

9)
$$
\begin{array}{r}
62 \\
\times \underline{\phantom{00}} \\
248
\end{array}
$$

10)
$$
\begin{array}{r}
6 \\
\times \underline{\phantom{00}} \\
36
\end{array}
$$

11)
$$
\begin{array}{r}
3 \\
\times \underline{\phantom{00}} \\
9
\end{array}
$$

12)
$$
\begin{array}{r}
\\
\times \underline{\phantom{00}} \\
\end{array}
$$

13)
$$
\begin{array}{r}
66 \\
\times \underline{\phantom{00}} \\
198
\end{array}
$$

14)
$$
\begin{array}{r}
70 \\
\times \underline{\phantom{00}} \\
140
\end{array}
$$

15)
$$
\begin{array}{r}
24 \\
\times \underline{\phantom{00}} \\
120
\end{array}
$$

16)
$$
\begin{array}{r}
2 \\
\times \underline{\phantom{00}} \\
16
\end{array}
$$

17)
$$
\begin{array}{r}
26 \\
\times \underline{\phantom{00}} \\
156
\end{array}
$$

18)
$$
\begin{array}{r}
80 \\
\times \underline{\phantom{00}} \\
160
\end{array}
$$

19)
$$
\begin{array}{r}
78 \\
\times \underline{\phantom{00}} \\
546
\end{array}
$$

20)
$$
\begin{array}{r}
5 \\
\times \underline{\phantom{00}} \\
45
\end{array}
$$

Missing numbers? No problem! Let's fill in the blanks and complete the puzzle!

1)
$$\begin{array}{r} 80 \\ \times \phantom{00} \\ \hline 320 \end{array}$$

2)
$$\begin{array}{r} 4 \\ \times \phantom{00} \\ \hline 20 \end{array}$$

3)
$$\begin{array}{r} 53 \\ \times \phantom{00} \\ \hline 318 \end{array}$$

4)
$$\begin{array}{r} 2 \\ \times \phantom{00} \\ \hline 12 \end{array}$$

5)
$$\begin{array}{r} 11 \\ \times \phantom{00} \\ \hline 33 \end{array}$$

6)
$$\begin{array}{r} 58 \\ \times \phantom{00} \\ \hline 464 \end{array}$$

7)
$$\begin{array}{r} 6 \\ \times \phantom{00} \\ \hline 36 \end{array}$$

8)
$$\begin{array}{r} 3 \\ \times \phantom{00} \\ \hline 25 \end{array}$$

9)
$$\begin{array}{r} 36 \\ \times \phantom{00} \\ \hline 144 \end{array}$$

10)
$$\begin{array}{r} 82 \\ \times \phantom{00} \\ \hline 656 \end{array}$$

11)
$$\begin{array}{r} 60 \\ \times \phantom{00} \\ \hline 240 \end{array}$$

12)
$$\begin{array}{r} 3 \\ \times \phantom{00} \\ \hline 34 \end{array}$$

13)
$$\begin{array}{r} 12 \\ \times \phantom{00} \\ \hline 60 \end{array}$$

14)
$$\begin{array}{r} 95 \\ \times \phantom{00} \\ \hline 475 \end{array}$$

15)
$$\begin{array}{r} 63 \\ \times \phantom{00} \\ \hline 126 \end{array}$$

16)
$$\begin{array}{r} 3 \\ \times \phantom{00} \\ \hline 24 \end{array}$$

17)
$$\begin{array}{r} 3 \\ \times \phantom{00} \\ \hline 15 \end{array}$$

18)
$$\begin{array}{r} 43 \\ \times \phantom{00} \\ \hline 387 \end{array}$$

19)
$$\begin{array}{r} 61 \\ \times \phantom{00} \\ \hline 366 \end{array}$$

20)
$$\begin{array}{r} 2 \\ \times \phantom{00} \\ \hline 16 \end{array}$$

  Time to solve some multiplication mysteries. Let's find those missing pieces!

 **Name:** _____   **Date:** _____

**Day 88** **Time Taken:** ___:___  **Score:** ___ / **20**

1)  37
    × ___
    185

2)  22
    × ___
    198

3)  93
    × ___
    279

4)    3
    × ___
    23

5)  33
    × ___
    66

6)  92
    × ___
    552

7)  63
    × ___
    504

8)    7
    × ___
    30

9)  20
    × ___
    160

10)  76
     × ___
     304

11)  10
     × ___
     30

12)   6
     × ___
     26

13)  25
     × ___
     225

14)  99
     × ___
     594

15)  22
     × ___
     176

16)   4
     × ___
     18

17)  67
     × ___
     335

18)  22
     × ___
     66

19)  23
     × ___
     46

20)   9
     × ___
     45

91

Missing numbers adding some intrigue? Let's uncover the secrets!

**Day 89**

 **Name:** _____

**Time Taken:** ___:___

 **Date:** _____

 **Score:** ___ / **20**

1)
$$\begin{array}{r} 56 \\ \times\ \underline{\phantom{00}} \\ 504 \end{array}$$

2)
$$\begin{array}{r} 88 \\ \times\ \underline{\phantom{00}} \\ 176 \end{array}$$

3)
$$\begin{array}{r} 17 \\ \times\ \underline{\phantom{00}} \\ 85 \end{array}$$

4)
$$\begin{array}{r} 9 \\ \times\ \underline{\phantom{00}} \\ 29 \end{array}$$

5)
$$\begin{array}{r} 42 \\ \times\ \underline{\phantom{00}} \\ 84 \end{array}$$

6)
$$\begin{array}{r} 19 \\ \times\ \underline{\phantom{00}} \\ 95 \end{array}$$

7)
$$\begin{array}{r} 12 \\ \times\ \underline{\phantom{00}} \\ 24 \end{array}$$

8)
$$\begin{array}{r} 5 \\ \times\ \underline{\phantom{00}} \\ 36 \end{array}$$

9)
$$\begin{array}{r} 11 \\ \times\ \underline{\phantom{00}} \\ 33 \end{array}$$

10)
$$\begin{array}{r} 31 \\ \times\ \underline{\phantom{00}} \\ 62 \end{array}$$

11)
$$\begin{array}{r} 96 \\ \times\ \underline{\phantom{00}} \\ 384 \end{array}$$

12)
$$\begin{array}{r} 5 \\ \times\ \underline{\phantom{00}} \\ 17 \end{array}$$

13)
$$\begin{array}{r} 60 \\ \times\ \underline{\phantom{00}} \\ 480 \end{array}$$

14)
$$\begin{array}{r} 62 \\ \times\ \underline{\phantom{00}} \\ 248 \end{array}$$

15)
$$\begin{array}{r} 76 \\ \times\ \underline{\phantom{00}} \\ 228 \end{array}$$

16)
$$\begin{array}{r} 9 \\ \times\ \underline{\phantom{00}} \\ 28 \end{array}$$

17)
$$\begin{array}{r} 61 \\ \times\ \underline{\phantom{00}} \\ 183 \end{array}$$

18)
$$\begin{array}{r} 10 \\ \times\ \underline{\phantom{00}} \\ 30 \end{array}$$

19)
$$\begin{array}{r} 97 \\ \times\ \underline{\phantom{00}} \\ 679 \end{array}$$

20)
$$\begin{array}{r} 1 \\ \times\ \underline{\phantom{00}} \\ 11 \end{array}$$

Let's add up the excitement by uncovering the missing numbers!

 **Name:** _____

 **Date:** _____

 **Time Taken:** ___:___

**Score:** / **20**

**Day 90**

1)
$$\begin{array}{r} 30 \\ \times \phantom{00} \\ \hline 270 \end{array}$$

2)
$$\begin{array}{r} 80 \\ \times \phantom{00} \\ \hline 400 \end{array}$$

3)
$$\begin{array}{r} 32 \\ \times \phantom{00} \\ \hline 160 \end{array}$$

4)
$$\begin{array}{r} 6 \\ \times \phantom{00} \\ \hline 13 \end{array}$$

5)
$$\begin{array}{r} 50 \\ \times \phantom{00} \\ \hline 350 \end{array}$$

6)
$$\begin{array}{r} 73 \\ \times \phantom{00} \\ \hline 146 \end{array}$$

7)
$$\begin{array}{r} 56 \\ \times \phantom{00} \\ \hline 504 \end{array}$$

8)
$$\begin{array}{r} 4 \\ \times \phantom{00} \\ \hline 21 \end{array}$$

9)
$$\begin{array}{r} 96 \\ \times \phantom{00} \\ \hline 864 \end{array}$$

10)
$$\begin{array}{r} 30 \\ \times \phantom{00} \\ \hline 60 \end{array}$$

11)
$$\begin{array}{r} 22 \\ \times \phantom{00} \\ \hline 176 \end{array}$$

12)
$$\begin{array}{r} 1 \\ \times \phantom{00} \\ \hline 11 \end{array}$$

13)
$$\begin{array}{r} 91 \\ \times \phantom{00} \\ \hline 819 \end{array}$$

14)
$$\begin{array}{r} 76 \\ \times \phantom{00} \\ \hline 608 \end{array}$$

15)
$$\begin{array}{r} 86 \\ \times \phantom{00} \\ \hline 430 \end{array}$$

16)
$$\begin{array}{r} 4 \\ \times \phantom{00} \\ \hline 9 \end{array}$$

17)
$$\begin{array}{r} 25 \\ \times \phantom{00} \\ \hline 200 \end{array}$$

18)
$$\begin{array}{r} 36 \\ \times \phantom{00} \\ \hline 180 \end{array}$$

19)
$$\begin{array}{r} 46 \\ \times \phantom{00} \\ \hline 184 \end{array}$$

20)
$$\begin{array}{r} 8 \\ \times \phantom{00} \\ \hline 34 \end{array}$$

 Let's divide and conquer today's math problems!

 Day 91

 Name: _____

Time Taken: ___:___

Date: _____

Score:  / 20

1) 
$$86 \div 2$$

2) 
$$6 \div 6$$

3) 
$$21 \div 3$$

4) 
$$3 \div$$

5) 
$$8 \div 2$$

6) 
$$74 \div 2$$

7) 
$$7 \div 7$$

8) 
$$6 \div$$

9) 
$$60 \div 4$$

10) 
$$25 \div 5$$

11) 
$$7 \div 7$$

12) 
$$9 \div$$

13) 
$$80 \div 4$$

14) 
$$88 \div 2$$

15) 
$$54 \div 6$$

16) 
$$9 \div$$

17) 
$$39 \div 3$$

18) 
$$76 \div 4$$

19) 
$$46 \div 2$$

20) 
$$2 \div$$

Division challenges await! Let's divide and conquer!

Day 92

 **Name:** _____

 **Date:** _____

**Time Taken:** ___:___

 **Score:** / **20**

1)
$$85 \div 5$$

2)
$$26 \div 2$$

3)
$$74 \div 2$$

4)
$$2 \div$$

5)
$$88 \div 2$$

6)
$$98 \div 2$$

7)
$$81 \div 9$$

8)
$$2 \div$$

9)
$$10 \div 5$$

10)
$$44 \div 4$$

11)
$$14 \div 7$$

12)
$$\div$$

13)
$$28 \div 2$$

14)
$$3 \div 3$$

15)
$$28 \div 4$$

16)
$$8 \div$$

17)
$$16 \div 8$$

18)
$$25 \div 5$$

19)
$$48 \div 6$$

20)
$$6 \div$$

95

  **Name:** _____    **Date:** _____

**Day 93**    **Time Taken:** ___:___    **Score:**    / 20

1)  51 ÷ 3

2)  78 ÷ 2

3)  64 ÷ 4

4)  2 ÷ ___

5)  85 ÷ 5

6)  20 ÷ 5

7)  35 ÷ 7

8)  9 ÷ ___

9)  78 ÷ 3

10)  74 ÷ 2

11)  8 ÷ 2

12)  1 ÷ ___

13)  12 ÷ 2

14)  80 ÷ 2

15)  57 ÷ 3

16)  4 ÷ ___

17)  3 ÷ 3

18)  93 ÷ 3

19)  96 ÷ 3

20)  3 ÷ ___

**Day 94**

 **Name:** _____     **Date:** _____

**Time Taken:** ___:___    **Score:**    / **20**

1)  45
    ÷  3

2)   3
    ÷  3

3)  14
    ÷  7

4)   9
    ÷

5)   5
    ÷  5

6)   2
    ÷  2

7)  24
    ÷  2

8)   5
    ÷

9)  33
    ÷  3

10)  35
     ÷  5

11)  46
     ÷  2

12)   6
     ÷

13)  82
     ÷  2

14)  28
     ÷  7

15)   6
     ÷  2

16)
     ÷

17)  76
     ÷  2

18)  65
     ÷  5

19)  20
     ÷  5

20)   4
     ÷

**Day 95**

 **Name:** _____

**Time Taken:** ___:___

 **Date:** _____

 **Score:** ___ / **20**

1)
$$48 \div 8$$

2)
$$84 \div 4$$

3)
$$24 \div 3$$

4)
$$5 \div$$

5)
$$6 \div 2$$

6)
$$98 \div 7$$

7)
$$57 \div 3$$

8)
$$3 \div$$

9)
$$56 \div 4$$

10)
$$93 \div 3$$

11)
$$52 \div 4$$

12)
$$5 \div$$

13)
$$18 \div 2$$

14)
$$70 \div 2$$

15)
$$52 \div 4$$

16)
$$8 \div$$

17)
$$77 \div 7$$

18)
$$24 \div 2$$

19)
$$24 \div 4$$

20)
$$2 \div$$

Let's activate our division powers. Division mode: on!

**Day 96**

 **Name:** _____

**Time Taken:** ____:____

 **Date:** _____

**Score:** ___ / 20

1)  78 ÷ 6

2)  22 ÷ 2

3)  46 ÷ 2

4)  7 ÷

5)  18 ÷ 2

6)  68 ÷ 2

7)  12 ÷ 3

8)  3 ÷

9)  72 ÷ 6

10)  46 ÷ 2

11)  84 ÷ 7

12)  3 ÷

13)  96 ÷ 4

14)  95 ÷ 5

15)  10 ÷ 2

16)  9 ÷

17)  64 ÷ 4

18)  26 ÷ 2

19)  22 ÷ 2

20)  5 ÷

Dividing numbers like a pro! Let's tackle these equations!

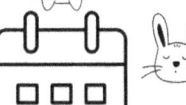

1)
$$72 \div 9$$

2)
$$98 \div 7$$

3)
$$36 \div 2$$

4)
$$9 \div$$

5)
$$68 \div 4$$

6)
$$18 \div 6$$

7)
$$72 \div 4$$

8)
$$\div$$

9)
$$27 \div 3$$

10)
$$90 \div 2$$

11)
$$69 \div 3$$

12)
$$8 \div$$

13)
$$54 \div 9$$

14)
$$60 \div 6$$

15)
$$95 \div 5$$

16)
$$1 \div$$

17)
$$90 \div 5$$

18)
$$45 \div 9$$

19)
$$77 \div 7$$

20)
$$1 \div$$

Ready to divide and simplify our math problems? Let's go!

**Day 98**

Name: _____

Time Taken: ____:____

Date: _____

Score: ____ / 20

1)   96
   ÷ 2

2)   49
   ÷ 7

3)   55
   ÷ 5

4)    9
   ÷

5)   50
   ÷ 5

6)   90
   ÷ 5

7)   45
   ÷ 5

8)    7
   ÷

9)   28
   ÷ 7

10)   70
    ÷ 7

11)   76
    ÷ 2

12)    5
    ÷

13)   93
    ÷ 3

14)   68
    ÷ 2

15)   80
    ÷ 5

16)    3
    ÷

17)   95
    ÷ 5

18)    6
    ÷ 2

19)   40
    ÷ 2

20)    6
    ÷

Let's divide and conquer the world of math today!

 **Name:** _____   **Date:** _____

 **Time Taken:** ___:___  **Score:** / 20

**Day 99**

1)  36
   ÷  6
   ____

2)  12
   ÷  3
   ____

3)   4
   ÷  2
   ____

4)   4
   ÷
   ____

5)  32
   ÷  2
   ____

6)  63
   ÷  7
   ____

7)  45
   ÷  3
   ____

8)   1
   ÷
   ____

9)   7
   ÷  7
   ____

10)  32
    ÷  2
    ____

11)  74
    ÷  2
    ____

12)   1
    ÷
    ____

13)   9
    ÷  3
    ____

14)  85
    ÷  5
    ____

15)  14
    ÷  7
    ____

16)   8
    ÷
    ____

17)  68
    ÷  4
    ____

18)  21
    ÷  7
    ____

19)  35
    ÷  7
    ____

20)   4
    ÷
    ____

Division manial Let's divide our way to success!

1)  45
    ÷ 9

2)  78
    ÷ 6

3)  18
    ÷ 2

4)  5
    ÷

5)  51
    ÷ 3

6)  25
    ÷ 5

7)  15
    ÷ 3

8)  4
    ÷

9)  5
    ÷ 5

10) 4
    ÷ 2

11) 84
    ÷ 2

12) 1
    ÷

13) 4
    ÷ 4

14) 5
    ÷ 5

15) 87
    ÷ 3

16) 9
    ÷

17) 93
    ÷ 3

18) 55
    ÷ 5

19) 54
    ÷ 6

20) 4
    ÷

# ANSWERS

 Hey buddy! Let's check if the answer you calculated is correct............. Come on!

**Day 1**

(1)73 (2)106 (3)153 (4)167 (5)89 (6)95
(7)67 (8)136 (9)93 (10)137 (11)122
(12)137 (13)22 (14)108 (15)135 (16)77
(17)107 (18)144 (19)100 (20)52

**Day 2**

(1)149 (2)17 (3)72 (4)80 (5)88 (6)170
(7)182 (8)112 (9)124 (10)97 (11)104
(12)134 (13)67 (14)102 (15)72 (16)35
(17)90 (18)107 (19)80 (20)165

**Day 3**

(1)70 (2)125 (3)67 (4)169 (5)50 (6)108
(7)107 (8)100 (9)145 (10)71 (11)108
(12)135 (13)83 (14)108 (15)112 (16)127
(17)92 (18)117 (19)76 (20)70

**Day 4**

(1)94 (2)137 (3)128 (4)130 (5)77 (6)34
(7)106 (8)37 (9)183 (10)95 (11)62 (12)148
(13)91 (14)71 (15)151 (16)163 (17)99
(18)48 (19)91 (20)31

**Day 5**

(1)132 (2)111 (3)121 (4)182 (5)143 (6)122
(7)136 (8)103 (9)63 (10)39 (11)39 (12)118
(13)142 (14)116 (15)116 (16)132 (17)143
(18)159 (19)167 (20)18

**Day 6**

(1)87 (2)100 (3)56 (4)37 (5)42 (6)120
(7)100 (8)67 (9)55 (10)61 (11)61 (12)43
(13)156 (14)103 (15)108 (16)98 (17)45
(18)74 (19)42 (20)116

**Day 7**

(1)133 (2)160 (3)100 (4)167 (5)116 (6)150
(7)131 (8)66 (9)114 (10)45 (11)98 (12)60
(13)39 (14)159 (15)100 (16)94 (17)155
(18)92 (19)127 (20)72

**Day 8**

(1)90 (2)106 (3)149 (4)74 (5)160 (6)75
(7)56 (8)94 (9)112 (10)106 (11)112 (12)65
(13)69 (14)29 (15)131 (16)162 (17)95
(18)80 (19)93 (20)124

**Day 9**

(1)104 (2)57 (3)92 (4)165 (5)86 (6)165
(7)103 (8)75 (9)75 (10)146 (11)104 (12)67
(13)90 (14)66 (15)89 (16)133 (17)99
(18)69 (19)45 (20)38

**Day 10**

(1)91 (2)143 (3)136 (4)87 (5)92 (6)85
(7)98 (8)93 (9)77 (10)143 (11)113 (12)123
(13)107 (14)80 (15)96 (16)36 (17)117
(18)138 (19)125 (20)50

# ANSWERS

Hey buddy! Let's check if the answer you calculated is correct........... Come on!

**Day 11**

(1)106 (2)87 (3)107 (4)38 (5)166 (6)101
(7)56 (8)157 (9)55 (10)71 (11)102 (12)129
(13)178 (14)73 (15)36 (16)139 (17)109
(18)177 (19)118 (20)117

**Day 12**

(1)67 (2)73 (3)105 (4)119 (5)140 (6)67
(7)171 (8)166 (9)72 (10)163 (11)155
(12)181 (13)184 (14)90 (15)104 (16)140
(17)64 (18)128 (19)167 (20)95

**Day 13**

(1)30 (2)80 (3)143 (4)82 (5)72 (6)161
(7)72 (8)107 (9)98 (10)61 (11)135 (12)111
(13)59 (14)77 (15)148 (16)84 (17)140
(18)43 (19)112 (20)132

**Day 14**

(1)159 (2)91 (3)93 (4)129 (5)107 (6)124
(7)40 (8)172 (9)127 (10)139 (11)111
(12)181 (13)116 (14)51 (15)133 (16)79
(17)144 (18)163 (19)119 (20)144

**Day 15**

(1)97 (2)100 (3)104 (4)65 (5)81 (6)100
(7)45 (8)159 (9)79 (10)46 (11)151 (12)121
(13)101 (14)152 (15)66 (16)88 (17)86
(18)67 (19)120 (20)91

**Day 16**

(1)146 (2)53 (3)58 (4)59 (5)104 (6)103
(7)148 (8)84 (9)93 (10)81 (11)43 (12)71
(13)46 (14)153 (15)127 (16)186 (17)59
(18)113 (19)175 (20)88

**Day 17**

(1)24 (2)129 (3)58 (4)99 (5)72 (6)86 (7)90
(8)108 (9)121 (10)77 (11)115 (12)155
(13)113 (14)132 (15)91 (16)59 (17)123
(18)117 (19)48 (20)109

**Day 18**

(1)40 (2)99 (3)140 (4)109 (5)112 (6)137
(7)107 (8)144 (9)116 (10)132 (11)158
(12)97 (13)67 (14)78 (15)152 (16)162
(17)178 (18)64 (19)89 (20)126

**Day 19**

(1)81 (2)96 (3)112 (4)86 (5)130 (6)103
(7)109 (8)119 (9)111 (10)78 (11)166
(12)99 (13)137 (14)76 (15)121 (16)106
(17)86 (18)173 (19)87 (20)89

**Day 20**

(1)131 (2)94 (3)60 (4)113 (5)97 (6)99
(7)64 (8)84 (9)114 (10)151 (11)78 (12)69
(13)170 (14)161 (15)109 (16)101 (17)103
(18)152 (19)94 (20)136

# ANSWERS

 Hey buddy! Let's check if the answer you calculated is correct............ Come on!

**Day 21**

(1)86 (2)79 (3)43 (4)61 (5)45 (6)84 (7)29
(8)16 (9)92 (10)68 (11)47 (12)72 (13)64
(14)88 (15)83 (16)51 (17)98 (18)77 (19)41
(20)20

**Day 22**

(1)27 (2)90 (3)32 (4)73 (5)26 (6)42 (7)91
(8)21 (9)87 (10)49 (11)96 (12)81 (13)11
(14)10 (15)11 (16)38 (17)7 (18)51 (19)95
(20)41

**Day 23**

(1)95 (2)99 (3)67 (4)96 (5)42 (6)20 (7)27
(8)98 (9)28 (10)1 (11)29 (12)65 (13)87
(14)99 (15)69 (16)64 (17)16 (18)43 (19)84
(20)63

**Day 24**

(1)20 (2)4 (3)50 (4)74 (5)73 (6)80 (7)53
(8)82 (9)13 (10)13 (11)48 (12)1 (13)96
(14)2 (15)12 (16)86 (17)46 (18)75 (19)82
(20)64

**Day 25**

(1)50 (2)17 (3)80 (4)89 (5)75 (6)21 (7)65
(8)92 (9)14 (10)53 (11)22 (12)22 (13)98
(14)93 (15)39 (16)78 (17)94 (18)45 (19)16
(20)27

**Day 26**

(1)30 (2)28 (3)90 (4)81 (5)82 (6)60 (7)93
(8)4 (9)37 (10)65 (11)91 (12)82 (13)45
(14)96 (15)75 (16)25 (17)35 (18)27 (19)50
(20)10

**Day 27**

(1)66 (2)28 (3)94 (4)18 (5)74 (6)4 (7)51
(8)55 (9)84 (10)82 (11)57 (12)10 (13)27
(14)87 (15)42 (16)98 (17)5 (18)29 (19)46
(20)13

**Day 28**

(1)77 (2)11 (3)22 (4)56 (5)45 (6)73 (7)65
(8)96 (9)30 (10)53 (11)86 (12)72 (13)65
(14)62 (15)97 (16)76 (17)91 (18)91 (19)90
(20)6

**Day 29**

(1)71 (2)50 (3)48 (4)30 (5)65 (6)97 (7)61
(8)87 (9)19 (10)33 (11)25 (12)92 (13)45
(14)49 (15)27 (16)36 (17)34 (18)39 (19)93
(20)38

**Day 30**

(1)22 (2)30 (3)86 (4)38 (5)55 (6)72 (7)36
(8)16 (9)41 (10)19 (11)15 (12)5 (13)3
(14)14 (15)81 (16)84 (17)33 (18)78 (19)19
(20)3

# ANSWERS

Hey buddy! Let's check if the answer you calculated is correct............... Come on!

**Day 31**

(1)44 (2)4 (3)27 (4)78 (5)19 (6)19 (7)41
(8)16 (9)18 (10)1 (11)26 (12)14 (13)34
(14)23 (15)23 (16)51 (17)36 (18)25 (19)8
(20)43

**Day 32**

(1)37 (2)37 (3)32 (4)71 (5)55 (6)0 (7)18
(8)20 (9)9 (10)15 (11)31 (12)41 (13)2
(14)37 (15)62 (16)24 (17)6 (18)83 (19)26
(20)28

**Day 33**

(1)26 (2)2 (3)52 (4)6 (5)63 (6)14 (7)8 (8)16
(9)16 (10)35 (11)4 (12)52 (13)32 (14)5
(15)3 (16)8 (17)38 (18)10 (19)35 (20)12

**Day 34**

(1)2 (2)41 (3)42 (4)36 (5)30 (6)26 (7)86
(8)49 (9)13 (10)66 (11)38 (12)1 (13)23
(14)32 (15)3 (16)10 (17)21 (18)67 (19)53
(20)59

**Day 35**

(1)8 (2)13 (3)67 (4)83 (5)25 (6)3 (7)16
(8)13 (9)31 (10)59 (11)17 (12)71 (13)35
(14)63 (15)72 (16)36 (17)55 (18)64 (19)26
(20)1

**Day 36**

(1)53 (2)8 (3)44 (4)23 (5)34 (6)6 (7)31 (8)6
(9)20 (10)47 (11)16 (12)32 (13)27 (14)76
(15)23 (16)21 (17)27 (18)25 (19)47 (20)51

**Day 37**

(1)14 (2)67 (3)52 (4)34 (5)31 (6)31 (7)62
(8)22 (9)15 (10)80 (11)63 (12)19 (13)7

(14)12 (15)4 (16)14 (17)9 (18)41 (19)27
(20)36

**Day 38**

(1)37 (2)8 (3)36 (4)33 (5)67 (6)9 (7)28
(8)30 (9)3 (10)20 (11)47 (12)38 (13)24
(14)51 (15)34 (16)0 (17)11 (18)55 (19)61
(20)3

**Day 39**

(1)21 (2)51 (3)46 (4)0 (5)44 (6)40 (7)59
(8)41 (9)22 (10)36 (11)28 (12)50 (13)4
(14)15 (15)1 (16)55 (17)48 (18)31 (19)14
(20)57

**Day 40**

(1)15 (2)14 (3)55 (4)25 (5)39 (6)42 (7)67
(8)0 (9)72 (10)29 (11)29 (12)44 (13)68
(14)25 (15)31 (16)32 (17)4 (18)57 (19)24
(20)54

**Day 41**

(1)32 (2)28 (3)37 (4)61 (5)31 (6)41 (7)21
(8)2 (9)32 (10)34 (11)69 (12)46 (13)4
(14)63 (15)73 (16)67 (17)38 (18)16 (19)48
(20)36

# ANSWERS

Hey buddy! Let's check if the answer you calculated is correct.............. Come on!

**Day 42**
(1)7 (2)51 (3)37 (4)25 (5)61 (6)3 (7)34
(8)80 (9)10 (10)2 (11)52 (12)9 (13)40
(14)49 (15)60 (16)33 (17)32 (18)16 (19)15
(20)51

**Day 43**
(1)24 (2)31 (3)9 (4)55 (5)7 (6)4 (7)30 (8)1
(9)26 (10)6 (11)37 (12)1 (13)41 (14)2
(15)67 (16)28 (17)25 (18)21 (19)33 (20)25

**Day 44**
(1)43 (2)18 (3)19 (4)65 (5)1 (6)30 (7)40
(8)46 (9)7 (10)52 (11)7 (12)35 (13)24
(14)86 (15)47 (16)13 (17)4 (18)34 (19)16
(20)29

**Day 45**
(1)19 (2)43 (3)24 (4)52 (5)73 (6)13 (7)38
(8)12 (9)35 (10)26 (11)44 (12)32 (13)56
(14)13 (15)44 (16)5 (17)49 (18)33 (19)20
(20)57

**Day 46**
(1)8 (2)42 (3)17 (4)35 (5)3 (6)11 (7)33 (8)1
(9)36 (10)23 (11)37 (12)25 (13)26 (14)24
(15)60 (16)5 (17)29 (18)79 (19)46 (20)65

**Day 47**
(1)33 (2)2 (3)38 (4)38 (5)28 (6)5 (7)0 (8)24
(9)42 (10)27 (11)18 (12)9 (13)11 (14)10
(15)1 (16)16 (17)0 (18)2 (19)26 (20)30

**Day 48**
(1)15 (2)0 (3)3 (4)9 (5)5 (6)23 (7)24 (8)31
(9)5 (10)37 (11)10 (12)65 (13)47 (14)55
(15)48 (16)9 (17)18 (18)24 (19)40 (20)25

**Day 49**
(1)77 (2)75 (3)63 (4)58 (5)19 (6)7 (7)14
(8)13 (9)29 (10)12 (11)13 (12)62 (13)17
(14)47 (15)48 (16)65 (17)78 (18)30 (19)22
(20)1

**Day 50**
(1)53 (2)36 (3)56 (4)56 (5)6 (6)28 (7)4
(8)15 (9)43 (10)47 (11)67 (12)16 (13)54
(14)9 (15)24 (16)32 (17)69 (18)0 (19)35
(20)27

**Day 51**
(1)2 (2)3 (3)30 (4)51 (5)75 (6)41 (7)42
(8)24 (9)60 (10)40 (11)67 (12)43 (13)30
(14)23 (15)77 (16)23 (17)23 (18)55 (19)23
(20)83

**Day 52**
(1)31 (2)16 (3)42 (4)4 (5)38 (6)34 (7)78
(8)20 (9)9 (10)36 (11)15 (12)21 (13)61
(14)19 (15)49 (16)39 (17)28 (18)18 (19)13
(20)11

**Day 53**
(1)1 (2)10 (3)42 (4)54 (5)35 (6)8 (7)11 (8)5
(9)37 (10)34 (11)6 (12)37 (13)9 (14)15

# ANSWERS

Hey buddy! Let's check if the answer you calculated is correct............ Come on!

(15)52 (16)28 (17)12 (18)67 (19)62 (20)47

**Day 54**

(1)4 (2)27 (3)52 (4)11 (5)65 (6)12 (7)8
(8)25 (9)4 (10)62 (11)6 (12)39 (13)8 (14)5
(15)71 (16)63 (17)17 (18)30 (19)12 (20)90

**Day 55**

(1)85 (2)1 (3)38 (4)8 (5)20 (6)19 (7)59
(8)26 (9)39 (10)53 (11)27 (12)15 (13)71
(14)12 (15)11 (16)8 (17)4 (18)11 (19)21
(20)5

**Day 56**

(1)10 (2)2 (3)43 (4)47 (5)44 (6)24 (7)19
(8)4 (9)74 (10)37 (11)34 (12)30 (13)17
(14)13 (15)2 (16)6 (17)21 (18)23 (19)25
(20)24

**Day 57**

(1)22 (2)53 (3)31 (4)1 (5)11 (6)11 (7)50
(8)78 (9)4 (10)38 (11)24 (12)5 (13)55
(14)18 (15)31 (16)70 (17)40 (18)7 (19)37
(20)64

**Day 58**

(1)58 (2)9 (3)26 (4)59 (5)22 (6)15 (7)2
(8)21 (9)92 (10)45 (11)9 (12)5 (13)68
(14)52 (15)71 (16)15 (17)15 (18)6 (19)8
(20)13

**Day 59**

(1)3 (2)24 (3)31 (4)20 (5)65 (6)3 (7)5 (8)57
(9)15 (10)28 (11)19 (12)15 (13)3 (14)9
(15)42 (16)33 (17)16 (18)63 (19)23 (20)36

**Day 60**

(1)89 (2)7 (3)49 (4)51 (5)62 (6)25 (7)80

(8)3 (9)35 (10)9 (11)4 (12)12 (13)2 (14)37
(15)18 (16)31 (17)16 (18)37 (19)31 (20)8

**Day 61**

(1)7209 (2)5187 (3)9120 (4)2914 (5)4602
(6)2108 (7)2716 (8)48 (9)2112 (10)300
(11)4700 (12)8370 (13)588 (14)4752
(15)297 (16)1344 (17)570 (18)2852
(19)364 (20)1691

**Day 62**

(1)6674 (2)1032 (3)6800 (4)3534 (5)4830
(6)4410 (7)3850 (8)1672 (9)6630 (10)2660
(11)570 (12)720 (13)504 (14)6080 (15)638
(16)2550 (17)3042 (18)175 (19)3233
(20)2240

**Day 63**

(1)2079 (2)1332 (3)676 (4)6336 (5)344
(6)230 (7)1716 (8)572 (9)4851 (10)715
(11)222 (12)2040 (13)135 (14)3168
(15)385 (16)7296 (17)1425 (18)3520
(19)8217 (20)1769

# ANSWERS

Hey buddy! Let's check if the answer you calculated is correct............ Come on!

**Day 64**

(1)592 (2)1936 (3)2255 (4)196 (5)1581
(6)4900 (7)3111 (8)4187 (9)3350 (10)1680
(11)3186 (12)3570 (13)2646 (14)3717
(15)1020 (16)4758 (17)64 (18)420
(19)4422 (20)2139

**Day 65**

(1)144 (2)2068 (3)1739 (4)3726 (5)4941
(6)4257 (7)4970 (8)2538 (9)810 (10)1428
(11)4380 (12)3082 (13)8051 (14)938
(15)1365 (16)429 (17)441 (18)3920
(19)7029 (20)5824

**Day 66**

(1)854 (2)365 (3)4914 (4)4420 (5)1840
(6)2668 (7)6 (8)192 (9)1400 (10)1000
(11)3569 (12)3690 (13)1104 (14)2070
(15)3807 (16)1411 (17)528 (18)4988
(19)5643 (20)2772

**Day 67**

(1)744 (2)1406 (3)3484 (4)195 (5)7938
(6)286 (7)1950 (8)2360 (9)4400 (10)4560
(11)7410 (12)2178 (13)747 (14)4770
(15)1591 (16)684 (17)5208 (18)9009
(19)1904 (20)4940

**Day 68**

(1)4930 (2)3404 (3)4 (4)92 (5)72 (6)1344
(7)900 (8)2278 (9)140 (10)444 (11)825
(12)1281 (13)2162 (14)1350 (15)4864
(16)2093 (17)2808 (18)790 (19)715
(20)3080

**Day 69**

(1)1225 (2)2200 (3)4214 (4)5040 (5)767
(6)1305 (7)26 (8)5244 (9)1296 (10)4032
(11)7920 (12)174 (13)5148 (14)770
(15)539 (16)3552 (17)900 (18)2688
(19)3477 (20)2847

**Day 70**

(1)171 (2)8835 (3)5159 (4)3360 (5)5980
(6)5589 (7)3672 (8)5775 (9)1316 (10)1298
(11)682 (12)2304 (13)5415 (14)612
(15)208 (16)3108 (17)2108 (18)7221
(19)5488 (20)4895

**Day 71**

(1)567 (2)28 (3)144 (4)430 (5)595 (6)336
(7)680 (8)165 (9)410 (10)485 (11)252
(12)608 (13)24 (14)100 (15)204 (16)354
(17)48 (18)200 (19)105 (20)27

**Day 72**

(1)62 (2)828 (3)24 (4)552 (5)553 (6)192
(7)118 (8)87 (9)204 (10)387 (11)189
(12)156 (13)440 (14)256 (15)216 (16)50
(17)144 (18)260 (19)216 (20)170

# ANSWERS

Hey buddy! Let's check if the answer you calculated is correct.............. Come on!

**Day 73**
(1)36 (2)488 (3)210 (4)260 (5)74 (6)215
(7)490 (8)75 (9)22 (10)329 (11)585
(12)485 (13)130 (14)518 (15)147 (16)448
(17)112 (18)672 (19)168 (20)560

**Day 74**
(1)288 (2)765 (3)332 (4)114 (5)201 (6)567
(7)12 (8)582 (9)553 (10)150 (11)819
(12)95 (13)576 (14)63 (15)693 (16)416
(17)234 (18)24 (19)237 (20)306

**Day 75**
(1)819 (2)28 (3)60 (4)16 (5)88 (6)264
(7)118 (8)69 (9)312 (10)342 (11)372
(12)235 (13)46 (14)522 (15)112 (16)315
(17)520 (18)264 (19)210 (20)114

**Day 76**
(1)324 (2)486 (3)504 (4)657 (5)162 (6)729
(7)92 (8)28 (9)136 (10)25 (11)474 (12)138
(13)196 (14)94 (15)456 (16)448 (17)441
(18)54 (19)90 (20)344

**Day 77**
(1)256 (2)156 (3)189 (4)259 (5)468 (6)390
(7)765 (8)260 (9)45 (10)384 (11)855
(12)312 (13)285 (14)192 (15)104 (16)144
(17)294 (18)310 (19)776 (20)568

**Day 78**
(1)304 (2)240 (3)357 (4)134 (5)184 (6)584
(7)192 (8)210 (9)168 (10)81 (11)320
(12)810 (13)273 (14)98 (15)160 (16)180
(17)102 (18)236 (19)216 (20)380

**Day 79**
(1)675 (2)266 (3)26 (4)112 (5)144 (6)315
(7)198 (8)32 (9)176 (10)567 (11)270
(12)35 (13)124 (14)576 (15)356 (16)534
(17)51 (18)116 (19)744 (20)28

**Day 80**
(1)112 (2)128 (3)384 (4)66 (5)609 (6)194
(7)873 (8)154 (9)696 (10)531 (11)50
(12)90 (13)122 (14)352 (15)371 (16)468
(17)148 (18)65 (19)486 (20)372

**Day 81**
(1)6 (2)6 (3)8 (4)2 (5)6 (6)2 (7)7 (8)8 (9)8
(10)7 (11)3 (12)6 (13)2 (14)5 (15)2 (16)4
(17)6 (18)2 (19)3 (20)4

**Day 82**
(1)5 (2)8 (3)8 (4)9 (5)8 (6)4 (7)3 (8)3 (9)7
(10)8 (11)3 (12)2 (13)6 (14)7 (15)5 (16)5
(17)3 (18)8 (19)8 (20)9

**Day 83**
(1)5 (2)8 (3)7 (4)6 (5)2 (6)7 (7)8 (8)3 (9)7
(10)5 (11)4 (12)9 (13)9 (14)6 (15)7 (16)3
(17)4 (18)5 (19)9 (20)4

# ANSWERS

 Hey buddy! Let's check if the answer you calculated is correct.............. Come on!

**Day 84**

(1)7 (2)5 (3)8 (4)8 (5)3 (6)3 (7)8 (8)6 (9)2 (10)8 (11)8 (12)3 (13)8 (14)5 (15)9 (16)2 (17)4 (18)8 (19)2 (20)3

**Day 85**

(1)4 (2)2 (3)5 (4)8 (5)3 (6)8 (7)3 (8)8 (9)8 (10)6 (11)5 (12)8 (13)6 (14)8 (15)3 (16)9 (17)7 (18)5 (19)7 (20)2

**Day 86**

(1)8 (2)8 (3)2 (4)8 (5)6 (6)5 (7)3 (8)3 (9)4 (10)6 (11)3 (12)4 (13)3 (14)2 (15)5 (16)7 (17)6 (18)2 (19)7 (20)8

**Day 87**

(1)4 (2)5 (3)6 (4)6 (5)3 (6)8 (7)6 (8)7 (9)4 (10)8 (11)4 (12)9 (13)5 (14)5 (15)2 (16)7 (17)5 (18)9 (19)6 (20)6

**Day 88**

(1)5 (2)9 (3)3 (4)7 (5)2 (6)6 (7)8 (8)4 (9)8 (10)4 (11)3 (12)4 (13)9 (14)6 (15)8 (16)4 (17)5 (18)3 (19)2 (20)5

**Day 89**

(1)9 (2)2 (3)5 (4)3 (5)2 (6)5 (7)2 (8)7 (9)3 (10)2 (11)4 (12)3 (13)8 (14)4 (15)3 (16)3 (17)3 (18)3 (19)7 (20)9

**Day 90**

(1)9 (2)5 (3)5 (4)2 (5)7 (6)2 (7)9 (8)5 (9)9 (10)2 (11)8 (12)7 (13)9 (14)8 (15)5 (16)2 (17)8 (18)5 (19)4 (20)4

**Day 91**

(1)43 (2)1 (3)7 (4)11 (5)4 (6)37 (7)1 (8)34 (9)15 (10)5 (11)1 (12)45 (13)20 (14)44

(15)9 (16)47 (17)13 (18)19 (19)23 (20)11

**Day 92**

(1)17 (2)13 (3)37 (4)8 (5)44 (6)49 (7)9 (8)10 (9)2 (10)11 (11)2 (12)1 (13)14 (14)1 (15)7 (16)29 (17)2 (18)5 (19)8 (20)34

**Day 93**

(1)17 (2)39 (3)16 (4)10 (5)17 (6)4 (7)5 (8)47 (9)26 (10)37 (11)4 (12)8 (13)6 (14)40 (15)19 (16)16 (17)1 (18)31 (19)32 (20)6

**Day 94**

(1)15 (2)1 (3)2 (4)33 (5)1 (6)1 (7)12 (8)11 (9)11 (10)7 (11)23 (12)9 (13)41 (14)4 (15)3 (16)1 (17)38 (18)13 (19)4 (20)21

**Day 95**

(1)6 (2)21 (3)8 (4)29 (5)3 (6)14 (7)19 (8)19 (9)14 (10)31 (11)13 (12)18 (13)9 (14)35 (15)13 (16)43 (17)11 (18)12 (19)6 (20)5

# ANSWERS

Hey buddy! Let's check if the answer you calculated is correct.............. Come on!

**Day 96**

(1)13 (2)11 (3)23 (4)14 (5)9 (6)34 (7)4
(8)11 (9)12 (10)23 (11)12 (12)18 (13)24
(14)19 (15)5 (16)47 (17)16 (18)13 (19)11
(20)6

**Day 97**

(1)8 (2)14 (3)18 (4)31 (5)17 (6)3 (7)18 (8)4
(9)9 (10)45 (11)23 (12)27 (13)6 (14)10
(15)19 (16)2 (17)18 (18)5 (19)11 (20)6

**Day 98**

(1)48 (2)7 (3)11 (4)23 (5)10 (6)18 (7)9
(8)37 (9)4 (10)10 (11)38 (12)18 (13)31
(14)34 (15)16 (16)17 (17)19 (18)3 (19)20
(20)9

**Day 99**

(1)6 (2)4 (3)2 (4)21 (5)16 (6)9 (7)15 (8)5
(9)1 (10)16 (11)37 (12)5 (13)3 (14)17
(15)2 (16)29 (17)17 (18)3 (19)5 (20)10

**Day 100**

(1)5 (2)13 (3)9 (4)19 (5)17 (6)5 (7)5 (8)8
(9)1 (10)2 (11)42 (12)5 (13)1 (14)1 (15)29
(16)19 (17)31 (18)11 (19)9 (20)9

www.ingramcontent.com/pod-product-compliance
Lightning Source LLC
Chambersburg PA
CBHW081108290526
45795CB00006B/2050